IMPROVING BUSINESS STATISTICS THROUGH INTERAGENCY DATA SHARING

SUMMARY OF A WORKSHOP

Caryn Kuebler and Christopher Mackie, Rapporteurs

Steering Committee for the Workshop on the Benefits of Interagency Business Data Sharing

Committee on National Statistics
Division of Behavioral and Social Sciences and Education

NATIONAL RESEARCH COUNCIL
OF THE NATIONAL ACADEMIES

THE NATIONAL ACADEMIES PRESS
Washington, D.C.
www.nap.edu

THE NATIONAL ACADEMIES PRESS 500 Fifth Street, NW Washington, DC 20001

NOTICE: The project that is the subject of this report was approved by the Governing Board of the National Research Council, whose members are drawn from the councils of the National Academy of Sciences, the National Academy of Engineering, and the Institute of Medicine. The members of the committee responsible for the report were chosen for their special competences and with regard for appropriate balance.

This study is supported by a contract between the National Academy of Sciences and the National Science Foundation (Number SBR-0112521). The work of the Committee on National Statistics is also provided by a consortium of federal agencies through the same grant from the National Science Foundation. Any opinions, findings, conclusions, or recommendations expressed in this publication are those of the author(s) and do not necessarily reflect the views of the organizations or agencies that provided support for the project.

Library of Congress Cataloging-in-Publication Data

Improving business statistics through interagency data sharing : summary of a workshop / Caryn Kuebler and Christopher Mackie, rapporteurs.
p. cm.
Includes bibliographical references.
ISBN 0-309-10261-8 (pbk.) — ISBN 0-309-66282-6 (pdfs) 1. United States—Statistical services—Congresses. 2. United States—Commerce—Statistics—Congresses. 3. Commercial statistics—United States—Congresses. I. Kuebler, Caryn. II. Mackie, Christopher D. III. National Research Council (U.S.). Committee on National Statistics.
HA37.U55I47 2006
650.072′7—dc22

2006023359

International Standard Book Number-13 978-0-309-10261-2

Additional copies of this report are available from The National Academies Press, 500 Fifth Street, NW, Lockbox 285, Washington, DC 20055; (800) 624-6242 or (202) 334-3313 (in the Washington metropolitan area); Internet, http://www.nap.edu

Printed in the United States of America

Suggested citation: National Research Council. (2006). *Improving Business Statistics Through Interagency Data Sharing: Summary of a Workshop*. Caryn Kuebler and Christopher Mackie, Rapporteurs. Steering Committee for the Workshop on the Benefits of Interagency Business Data Sharing. Committee on National Statistics, Division of Behavioral and Social Sciences and Education. Washington, DC: The National Academies Press.

THE NATIONAL ACADEMIES

Advisers to the Nation on Science, Engineering, and Medicine

The **National Academy of Sciences** is a private, nonprofit, self-perpetuating society of distinguished scholars engaged in scientific and engineering research, dedicated to the furtherance of science and technology and to their use for the general welfare. Upon the authority of the charter granted to it by the Congress in 1863, the Academy has a mandate that requires it to advise the federal government on scientific and technical matters. Dr. Ralph J. Cicerone is president of the National Academy of Sciences.

The **National Academy of Engineering** was established in 1964, under the charter of the National Academy of Sciences, as a parallel organization of outstanding engineers. It is autonomous in its administration and in the selection of its members, sharing with the National Academy of Sciences the responsibility for advising the federal government. The National Academy of Engineering also sponsors engineering programs aimed at meeting national needs, encourages education and research, and recognizes the superior achievements of engineers. Dr. Wm. A. Wulf is president of the National Academy of Engineering.

The **Institute of Medicine** was established in 1970 by the National Academy of Sciences to secure the services of eminent members of appropriate professions in the examination of policy matters pertaining to the health of the public. The Institute acts under the responsibility given to the National Academy of Sciences by its congressional charter to be an adviser to the federal government and, upon its own initiative, to identify issues of medical care, research, and education. Dr. Harvey V. Fineberg is president of the Institute of Medicine.

The **National Research Council** was organized by the National Academy of Sciences in 1916 to associate the broad community of science and technology with the Academy's purposes of furthering knowledge and advising the federal government. Functioning in accordance with general policies determined by the Academy, the Council has become the principal operating agency of both the National Academy of Sciences and the National Academy of Engineering in providing services to the government, the public, and the scientific and engineering communities. The Council is administered jointly by both Academies and the Institute of Medicine. Dr. Ralph J. Cicerone and Dr. Wm. A. Wulf are chair and vice chair, respectively, of the National Research Council.

www.national-academies.org

STEERING COMMITTEE FOR THE WORKSHOP ON THE BENEFITS OF INTERAGENCY BUSINESS DATA SHARING

CHARLES L. SCHULTZE (*Chair*), The Brookings Institution, Washington, DC
DANIEL R. FEENBERG, National Bureau of Economic Research, Cambridge, MA
JOHN HALTIWANGER, Department of Economics, University of Maryland
RALPH A. RECTOR, The Heritage Foundation, Washington, DC

CHRISTOPHER MACKIE, *Study Director*
CARYN KUEBLER, *Research Associate*
MICHAEL SIRI, *Senior Program Assistant*

COMMITTEE ON NATIONAL STATISTICS
2005-2006

Preface

The workshop summarized in this report was convened by the Committee on National Statistics (CNSTAT) on behalf of the Bureau of Economic Analysis (BEA) to discuss interagency business data sharing. Recent legislation, particularly the 2002 Confidential Information Protection and Statistical Efficiency Act (CIPSEA), has created new opportunities for sharing among BEA, the Census Bureau, and the Bureau of Labor Statistics (BLS) and has also stirred debate on data sharing, access, and confidentiality issues. The purpose of this workshop was to present strategies for enhancing the ability of statistical agencies to efficiently share administrative and statistical data on businesses, while still protecting respondent confidentiality. More effective interagency data sharing is essential if the statistical agencies are to have access to the most accurate information available and, in turn, continue to improve the quality of data and statistics in a cost-effective and minimally burdensome manner.

The workshop focused on the benefits of data sharing to two groups of stakeholders—the statistical agencies themselves and downstream data users, such as researchers and policy makers. Presenters represented four key agencies involved in business data sharing—BEA, the Census Bureau, BLS, and the Internal Revenue Service (IRS)—as well as the Federal Reserve Board, the Congressional Budget Office, and academia. This report provides a summary of the presentations and the discussions that took place. The workshop was not designed to produce recommendations; however, participants persuasively presented arguments in favor of expanding data sharing—emphasizing increased efficiency, reduced respondent burden, and more accurate information for policy makers—and suggestions on ways to work within (and to change, if necessary) current codes and regulation to make it happen. Workshop presenters and participants also recognized the importance of maintaining high standards for protecting data confidentiality. Steven Landefeld, director of BEA, suggested implementing incremental changes to data-sharing arrangements, including streamlining administrative procedures under CIPSEA; expediting access to research data centers; and modifying IRS procedures, through legislative or regulatory changes, to promote effective use of administrative data for statistical uses.

On behalf of CNSTAT, I thank all of the workshop participants, particularly those who prepared detailed presentations, which provided for engaging and productive discussion. We also thank the workshop participants for their insightful comments and fruitful exchange of ideas, as well as for their input as staff drafted this report. We especially thank members of the workshop steering committee—Daniel Feenberg of the National Bureau of Economic Research, John Haltiwanger of the University of Maryland, and Ralph Rector of the Heritage Foundation—for their helpful guidance and leadership in planning and moderating the workshop. Robert Parker provided important consultation to the staff and the committee. We thank him, along with Dennis Fixler, Steven Landefeld, Nick Greenia, Mark Mazur, and George Plesko, for contributing important background papers for the workshop and this volume.

We are grateful for BEA's sponsorship of the workshop and thank Steven Landefeld and Dennis Fixler, in particular, for offering guidance to staff and the workshop steering committee in development of the agenda and in identifying the workshop goals. We would also like to thank the Kauffman Foundation, which is sponsoring a CNSTAT study on improving business data and statistics, whose members provided additional guidance for the development of the workshop program. Christopher Mackie, the staff study director for the workshop, was instrumental in every aspect of the workshop process. He stewarded the planning of the workshop, worked closely with the steering committee and participants, and, in collaboration with Caryn Kuebler, research associate, prepared and revised the report on the basis of comments from reviewers and workshop participants. We thank Michael Siri for expertly managing all the administrative details and workshop arrangements and for his work on the report itself. Christine McShane's technical editing substantially improved the report's readability. Connie Citro, director of CNSTAT, provided guidance and support throughout the project. Kirsten Sampson Snyder guided this report through the review process.

This report has been reviewed in draft form by individuals chosen for their diverse perspectives and technical expertise, in accordance with procedures approved by the Report Review Committee of the National Research Council. The purpose of this independent review is to provide candid and critical comments that assist the institution in making the published report as sound as possible and to ensure that the report meets institutional standards for objectivity, evidence, and responsiveness to the study charge. The review comments and draft manuscript remain confidential to protect the integrity of the deliberative process.

We wish to thank the following individuals for their participation in the review of this report: Katharine G. Abraham, Joint Program in Survey Methodology, University of Maryland; Daniel R. Feenberg, Information and Research Systems, National Bureau of Economic Research, Cambridge, MA; and Sanders Korenman, School of Public Affairs, Baruch College of the City University of New York.

The review of this report was overseen by Julia Lane, National Opinion Research Center and the University of Chicago. Although the individuals listed above have provided constructive comments and suggestions, responsibility for the final content of this report rests entirely with the National Academies.

Charles Schultze, *Chair*
Steering Committee for the Workshop on the
Benefits of Interagency Business Data Sharing

Contents

PART I
WORKSHOP SUMMARY

PART II
BACKGROUND PAPERS

Part I

Workshop Summary

1

Introduction

BACKGROUND

U.S. business data are used broadly, providing the building blocks for key national—as well as regional and local—statistics measuring aggregate income and output, employment, investment, prices, and productivity. Beyond aggregate statistics, individual- and firm-level data are used for a wide range of microanalyses by academic researchers and by policy makers. In the United States, data collection and production efforts are conducted by a decentralized system of statistical agencies.[1] This apparatus yields an extensive array of data that, particularly when made available in the form of microdata, provides an unparalleled resource for policy analysis and research on social issues and for the production of economic statistics. However, the decentralized nature of the statistical system also creates challenges to efficient data collection, to containment of respondent burden, and to maintaining consistency of terms and units of measurement. It is these challenges that raise to paramount importance the practice of effective data sharing among the statistical agencies.

During the workshop's introductory session, Steven Landefeld—director of the Bureau of Economic Analysis (BEA), the workshop's sponsoring agency—provided an overview of the goals motivating the event. He reflected on issues that arise in a decentralized statistical system, noting that its data products excel in detail, timeliness, and relevance but

[1]See Norwood (1995) for an account of this historical development.

often lag behind in consistency. These inconsistencies create problems for BEA in producing the national income and product accounts, as they must draw from numerous data sources and make adjustments for differences in collection timing, as well as in concepts and definitions. The quality of data produced by the statistical agencies, in turn, affects the work of users, including other agencies, such as those responsible for budget projections and planning, the allocation of funds, and state and local decision making. Landefeld pointed out that, while data sharing has already improved and facilitated the work of BEA, current arrangements are limited in key ways. For example, the codes and regulations of the Internal Revenue Service (IRS) allow the Census Bureau to share data on large multiestablishment businesses but, for reasons discussed below, not on smaller and single-establishment businesses, which account for about 40 percent of receipts nationwide. As a result, critical data omissions persist as BEA and the Bureau of Labor Statistics (BLS) go about the business of producing statistical information on various dimensions of the U.S. economy.

The stated purpose of the workshop, described in this summary report, was to present ideas for easing constraints limiting the ability of statistical agencies to efficiently share administrative and statistical data on businesses. In order to produce the highest quality data sets and statistics at the lowest possible cost—and with minimal respondent burden—statistical agencies must be able to access the best information available, system-wide.

With this as the backdrop, BEA asked the Committee on National Statistics of the National Academies to convene a workshop to discuss interagency business data sharing. The workshop was held October 21, 2005. Recent legislation, most notably the 2002 Confidential Information Protection and Statistical Efficiency Act (CIPSEA), has served to revive debate on data-sharing, access, and confidentiality issues. Although U.S. statistical agencies have a long history of data sharing and of efforts to improve those arrangements, CIPSEA has created new opportunities to expand interagency sharing of business data among BEA, the Census Bureau, and BLS.[2]

The CIPSEA legislation embodies two core goals: to establish uniform cross-agency confidentiality protections and to promote efficiency in the production of the nation's statistics by authorizing limited sharing of busi-

[2]Chapter 5 provides a detailed description of the history of data sharing and data-sharing legislation. In addition, Appendix B provides brief summaries of relevant data-sharing legislation.

ness data for statistical purposes. The objectives behind the data-sharing component of the legislation are threefold. First, it was hoped that permitting the three agencies to share information would improve the comparability and accuracy of federal statistics by allowing more timely updating of sample frames, development of consistent classifications of establishments and industries, and exploitation of administrative data. Second, more integrated use of data should reduce the paperwork burden for surveyed businesses. Finally, through these mechanisms, it was hoped that the sharing of data would lead to improved understanding of the U.S. economy, especially for key industry and regional statistics.

One example of CIPSEA's potential is reflected in the recently authorized ability of the Census Bureau and BEA to link survey data to produce new statistics on domestic and international U.S. research and development activity. As Katherine Wallman, the chief statistician of the Office of Management and Budget (OMB), pointed out, input from the Census Bureau, BEA, and BLS is essential if CIPSEA implementation and guidance are to successfully build on experiences from earlier data sharing to make future arrangements more effective. Wallman noted that at least some additional access to tax information will be needed in order to realize the full benefit of the umbrella legislation for data sharing among the three statistical agencies.

Tax data have always been an essential, but highly restricted, source of information for measuring aspects of the economy in general and for construction of the national income and product accounts in particular. Since long before CIPSEA, BEA has been able to utilize, in a limited manner, valuable business tax and revenue data from the IRS. For example, provisions in Section 6103 of the IRS code authorize BEA to access corporate income tax return information so that published IRS corporate profits data can be converted to accounting concepts appropriate for use in measuring gross domestic product. For its regional economic accounts program, BEA has been authorized under other provisions of Section 6103 to review individual tax return records in order to produce tabulations of nonfarm proprietors' income (which are reviewed by the IRS to ensure taxpayer confidentiality); these estimates are used, in turn, to distribute BEA's national totals by state and county.[3]

The above uses notwithstanding, BEA access to federal tax information is still extremely limited relative to that afforded to the Census Bureau. The current tax code allows the IRS to supply enough information (e.g., names and addresses) from businesses' tax returns so that the Cen-

---

[3]For more examples of data sharing, see Chapter 5.

sus Bureau can construct its business register and sampling frames; however, "commingled" Census Bureau-IRS data cannot be shared with either BEA or BLS. Because of the lack of specific legislative authority, Title 13 and IRS code (Title 26) guidelines vary with regard to whether or not the data collected by the Census Bureau directly from taxpayers (using the IRS-based sampling frame) are fully under their authority or whether the IRS should maintain some control. Tax data issues received considerable attention at the workshop and are reported on more fully below.

Before moving on to describe the workshop proceedings, it is useful to clarify a few terms that are used throughout this summary:[4]

- *Data sharing* is the exchange of information collected from businesses and individuals or reported to the IRS in identifiable form for statistical purposes.
- *Business data* include operating, financial, and related information about businesses, tax exempt organizations, and government entities (CIPSEA).
- *Identifiable form* means information that permits the identity of the respondent to whom the information applies to be reasonably inferred by either direct or indirect means.
- *Statistical purposes* involve the description, estimation, or analysis of the characteristics of groups, without identifying the individuals or organizations that comprise such groups. The designation also includes methods and procedures related to the collection, compilation, processing, or analysis of data about these groups and the development of related measurement methods, models, statistical classifications, or sampling frames.

Box 1-1 lists acronyms and abbreviations related to interagency business data sharing.

WORKSHOP CONTENT

The workshop focused on the benefits of data sharing to two groups of stakeholders: the statistical agencies themselves and downstream data users. Presenters were asked to highlight untapped opportunities for productive data sharing that cannot yet be exploited because of regulatory or legislative constraints. The most prominently discussed example was that of tax data needed to reconcile the two primary business lists used

[4]These definitions are expanded upon in Chapter 5.

BOX 1-1
Abbreviations and Acronyms Related to Interagency Business Data Sharing

BEA	Bureau of Economic Analysis
BLS	Bureau of Labor Statistics
BR	Business Register
CBO	Congressional Budget Office
CBP	County Business Patterns
CES	Current Employment Statistics
CIPSEA	Confidential Information Protection and Statistical Efficiency Act (2002)
COS	Company Organization Survey
CPS	Current Population Survey
EIN	Employer Identification Number
FASB	Financial Accounting Standards Board
FRB	Federal Reserve Board
FTI	Federal Tax Information
GAAP	Generally Accepted Accounting Principles
GAO	Government Accountability Office
GDI	Gross Domestic Income
GDP	Gross Domestic Product
IRC	Internal Revenue Code
IRS	Internal Revenue Service
JCT	Joint Committee on Taxation
MWR	Multiple Worksite Report
NAICS	North American Industry Classification System
NABE	National Association for Business Economics
OMB	Office of Management and Budget
PEO	Professional Employer Organization
QCEW	Quarterly Census of Employment and Wages
RDC	Research Data Center
SBO	Survey of Business Ownership
SIC	Standard Industrial Classification
SIRD	Survey of Industrial Research and Development
SOI	Statistics of Income
SSA	Social Security Administration
SSN	Social Security Number
SSS	Special Sworn Status
TIN	Taxpayer Identification Number
USDA-NASS	U.S. Department of Agriculture-National Agricultural Statistics Service

by the statistical agencies. Both BLS and the Census Bureau compile business establishment lists—the Business Establishment List and the Business Register, respectively—mainly from administrative data, but also with supplemental survey data. Each covers about 8 million business establishments with employees, and they are used for similar purposes: to create sampling frames for a wide variety of surveys by the Census Bureau and by other statistical agencies, for benchmarking survey data, for publishing employment and wage data, and for generating aggregates used by other agencies, most notably many of the inputs to the national income and product accounts (Box 1-2).

In addition to leading to discrepancies of coverage, the redundancy of effort creates inefficiencies in maintaining up-to-date frames and samples, and it may contribute to difficulties in achieving adequate response rates to various surveys. Combining information from both sources could generate a more accurate, consistent business list that, for BEA, could improve its estimates in a number of areas (e.g., trade in services, corporate profits and industry employment, and wages by location). Inconsistencies, particularly in the assignment of establishments to industries and the range of entities covered, carry direct implications for the reliability of key statistics—from gross domestic product, to employment, to produc-

BOX 1-2
Why Should BLS and the Census Bureau Work Toward a Reconciled Sampling Frame?

The argument for the business list case goes beyond simply reducing redundancy and, possibly, administrative expenses. As noted by several of the workshop's presenters, some widely used macro statistics are derived from combinations of the Census Bureau and BLS data. For example, productivity is calculated as the ratio of a Census Bureau figure (output) and a BLS figure (labor input). If output and input measures were estimated from the same survey, then the presence or absence of any particular firm from the sample would likely have a very small effect on the ratio, because both the numerator and the denominator would change in the same direction. With separate samples, however, even a relatively common occurrence, such as a discrepancy in the industry code of a firm, could have visible effects on aggregate industry productivity growth in two industries. This is because the change in output of the firm is attributed to one industry while the change in inputs is attributed to another. There are important research and policy incentives for moving toward the use of a common sampling frame.

tivity and industrial production—derived from business lists or subsequent survey-based data. Streamlining the business registers and survey programs is also likely to reduce respondent burden for businesses. In its presentation, BEA cited these and other improvements that could be achieved through more extensive integration of data across statistical agencies. However, it is important to note that there are advantages to having separate lists, as it allows each agency to tailor the characteristics of its list specifically to the purpose it serves.

Jim Spletzer and Paul Hanczaryk (presenting for BLS and the Census Bureau, respectively) noted that interagency data sharing is an obvious and low-cost prerequisite for improving the business lists. While list comparison work is well under way, the idea of a business list reconciliation project is still very much at the discussion stage. The legalities and procedures necessary to begin this kind of work (specifically, the restrictions resulting from the federal tax information that the Census Bureau receives from the IRS) are not trivial. If sharing among the three CIPSEA-designated agencies is to be fully exploited, either IRS regulations or code (or both) must be changed. It was known at the time of its passage that, in order for the Census Business Register to be shared, companion legislation to CIPSEA would be needed to modify Section 6103 of the IRS tax code or to change interpretation of that code. The Joint Committee on Taxation has not yet taken action to address this specific data-sharing need. The 2004 *Statistical Programs of the United States Government* report (Office of Management and Budget, 2004) indicated that the proposal for companion legislation, which would make complementary changes to the provisions in the "statistical use" section of the IRS code, was endorsed by the Treasury Department and submitted to Congress; however, it expired with the 107th Congress.

During the workshop's morning session, representatives from BLS, the Census Bureau, the IRS, and BEA addressed current data-sharing arrangements and the role that data sharing plays in producing federal statistics. The Census Bureau and BLS provided information on their ongoing business list comparison project, which is intended to comprehensively document the comparability of the lists. Mark Mazur and Nick Greenia of the IRS Research, Analysis, and Statistics Division provided an overview of current data-sharing arrangements and interpretation of relevant regulations and legislation. They expressed a clear understanding of the importance of data sharing for purposes of improving business lists and indicated a willingness to work carefully and incrementally toward this goal within the legal guidelines. They further suggested that, if companion legislation to CIPSEA is to have a real chance of moving forward, the expansion of tax data sharing should be narrow in scope and clearly tied to purpose. For example, for the purpose of reconciling business lists, perhaps only a few basic variables—e.g., name, address (or geocode), employer identification number, employment, payroll, industry—would need to be shared.

During the afternoon session, which highlighted the perspectives of downstream data users, presenters from the Federal Reserve Board (FRB), the Congressional Budget Office, and academia discussed the benefits of business data sharing as it relates to productivity and real output measurement, informing monetary policy, estimating business profits, and budget forecasting. Federal Reserve policy models incorporate productivity statistics that are derived from industry-level output and employment data. Dennis Fixler of BEA and Carol Corrado of FRB described how the maintenance of two establishment lists creates time-consuming complications. For example, for calculating industry productivity statistics, inconsistencies arise because source data are drawn using different methods and from nonidentical sets of business entities. Output figures (the numerator) originate from the Census Bureau data, while input figures (the denominator) are derived from BLS data.

Several participants touched on the distinction between informative data discrepancies and those leading to statistical inconsistencies that are costly in terms of user communities' time and resources. For example, although household and payroll-based estimates of employment differ significantly at times, the two sources can illuminate slightly different aspects of the labor market picture. And, although BEA would certainly like to minimize the statistical discrepancy between income- and expenditure-side measures of gross domestic product (particularly for a few problematic industries), employer and household surveys each generate valuable information, both independently and in combination. In contrast, reconciliation of the two business lists involves mainly definitional and classification issues which, workshop presenters seemed to agree, should be as consistent as possible.

Carol Corrado, chief of the Industrial Output Section in the FRB Division of Research and Statistics, noted that discrepancies between the Census Bureau and BLS employment by industry figures lessen her confidence in BEA's industry accounts and, in turn, the accuracy of productivity change measures. She added that it is likely that the sectors of the economy experiencing large changes in productivity may also be those associated with problematic data inconsistencies. Corrado argued further that the statistical agencies could use resources more efficiently if they did not have to maintain two business lists. Steering committee member John Haltiwanger noted that, given the different uses of the business lists, it would not make sense to choose one over another; instead the weaknesses and strengths of each should be recognized and exploited. Corrado expressed the hope that, in the very near future, a system would be in place that is capable of reconciling differences in employment by industry.

Dale Jorgenson of Harvard University touched on similar themes in his presentation, observing that policy makers are hamstrung by the fact that the data system exhibits inconsistencies that arise because of the absence of statutory authority to share data among agencies. He expressed the view that resolving these uncertainties is essential and suggested beginning at the most fundamental level. One goal of Jorgenson's work on a "new architecture" for the national income and product accounts is to have common registers of firms, establishments, families, and individuals and to collect the data in a way that is internally consistent at the micro level (Jorgenson, Landefeld, and Nordhaus, 2006).

Several presenters suggested that an effective approach to advancing the dialogue between policy makers and statistical agencies is to begin by recognizing the potential, not just for increased efficiency and more accurate information for policy, but also for reducing respondent burden. The logic behind the change must include compelling data-driven cases for which the payoff is clear across data-sharing agencies and between respondents and the agencies. In that spirit, workshop participants cited numerous examples in which more effective sharing would improve data, and for which the associated benefits more than warrant action to build on the successful data sharing authorized by CIPSEA.

Finally, the confidentiality side of CIPSEA was not neglected. Participants from the agencies stated the need to continue to take this responsibility very seriously as a matter of principle and as a means to buttress public confidence in the agencies. As outlined in the summary that follows, many presenters argued that the uniform confidentiality provisions created under CIPSEA provide sufficient coverage to expand data-sharing arrangements while still ensuring that the privacy and confidentiality of records will be maintained. At the time of the workshop, however, the confidentiality provisions under CIPSEA did not have OMB guidelines, and agencies continued to interpret the requirements differently. The point was made that, given the confidentiality requirements enacted through CIPSEA, the agencies are now in a better position than ever before to protect the data collected for statistical purposes under a pledge of confidentiality. Landefeld noted the importance of continuing to do a good job of protecting the confidentiality of data, while suggesting incremental changes in data-sharing arrangements, including:

- streamlining administrative procedures under CIPSEA;
- expediting access to research data centers (keeping statistical uses as top priority); and

- modifying IRS procedures to promote effective use of administrative data for statistical uses, either through legislation or changes in the regulations.

Landefeld expressed a hope that participants at the workshop would emerge with a renewed sense of the importance of moving forward to responsibly expand interagency data-sharing arrangements.

2

The Benefits of Data Sharing to the Statistical Agencies

Representatives from the Bureau of Labor Statistics (BLS), the Census Bureau, the Internal Revenue Service (IRS), and the Bureau of Economic Analysis (BEA) gave presentations on the benefits of data sharing to statistical agencies and on the prospects of enhancing current arrangements. Throughout the session, discussion about the data underlying the BLS and Census Bureau business lists was prominent. BLS and the Census Bureau are currently jointly engaged in a business list comparison project; Jim Spletzer and Paul Hanczaryk presented preliminary results and recommended next steps. Mark Mazur and Nick Greenia provided an explanation of current data-sharing arrangements and legal constraints from the IRS perspective. They also commented on the value of tax information to the statistical agencies and discussed the viability of various strategies for dealing with current data-sharing restrictions. Dennis Fixler delivered the morning's final presentation, an overview of the current role and future potential of data sharing to serve BEA's national income and product accounts work.

BUSINESS LIST COMPARISON AND RECONCILIATION

Both BLS and the Census Bureau compile business establishment lists—the Business Establishment List (BEL) and the Business Register (BR), respectively—mainly from administrative records, but also supplemented with survey data. Each "register" covers about 8 million business employer establishments, and they are used for similar purposes: to create sampling frames, to benchmark survey data, to publish employment and wage data, and to provide aggregate measures to other agencies. Products of the BEL and the BR include the Quarterly Census of Employment and Wages (QCEW) and the County Business Patterns (CBP), respectively. Within each agency's business list programs, there are numerous tasks for which data sharing could be helpful. For example, both the Census Bureau and BLS require establishment-level data for multiunit firms. The Census Bureau requests that these firms break out employment and payroll numbers by establishment in its Company Organization Survey (COS); for BLS, the Unemployment Insurance program's Multiple Worksite Report (MWR) is used. Because the timeliness and comprehensiveness of the COS and the MWR are not the same, combining results could enhance the measurement of employment, payroll, and establishment birth and death trends for multiunit firms.

Spletzer described the collaborative project ongoing at BLS and the Census Bureau to compare, improve, and (perhaps eventually) reconcile the two lists. The goals of the comparison project are twofold: to understand the differences in the lists and to identify the strengths and weaknesses of each. Contributors to the project at BLS and the Census Bureau are ultimately motivated by the prospect of identifying opportunities to improve the value of the lists in the context of the uses that they serve. Improving the comparability and accuracy of the Census Bureau and BLS business lists not only would provide benefits to the statistical agencies and to downstream users, but also could reduce reporting burden on the business community and possibly reduce costs to the agencies in the long run. Preliminary comparisons have found that heterogeneity between lists increases at finer levels of industry and geography detail; thus Spletzer and Hanczaryk suggest the need for greater sharing of micro-level data. Additionally, they noted the work the agencies plan to do to resolve the different methods used to determine single versus multi-establishment status.

One of the purposes of the Confidential Information Protection and Statistical Efficiency Act of 2002 (CIPSEA) is "to improve the comparability and accuracy of Federal economic statistics by allowing the Bureau of the Census, the Bureau of Economic Analysis, and the Bureau of Labor Statistics to update sample frames, develop consistent classifications of establishments and companies into industries, improve coverage, and reconcile significant differences in data produced by the three agencies" (Public Law 107–347, Subtitle B—Statistical Efficiency, Sec. 521, Findings and Purposes). The main hurdle to business list coordination work that arises involves the statutory restrictions on federal tax information. The bulk of the data underlying the Census Bureau register originates from IRS tax records and is shared under Title 26; however, Title 26 does not authorize BLS or BEA to access these records or, for that matter, the Cen-

TABLE 2-1 Comparison of Published Statistics

2001 Data	County Business Patterns	Quarterly Census of Employment and Wages	% Difference
Establishments	7,095,302	7,213,611	–1.7
Employment	115,061,184	108,916,710	5.5
Payroll (millions)	3,989,086	3,972,605	0.4

NOTE: Figures are adjusted for differences in industry coverage.

sus Bureau data commingled with them. Therefore, change in the tax code is required before microdata from IRS sources can be shared for programmatic purposes.

To inform the business list comparison project, BLS and Census Bureau analysts first evaluated and compared the published aggregate statistics. In order to make a comparison, adjustments were made to take into account differences in the scope and coverage of the BR and the QCEW, most notably to account for the fact that the former includes large segments of the self-employed business population that the latter does not.[1] BLS private-sector data needed filtering to remove certain industries that the Census Bureau does not cover, such as crop and animal production, rail transportation, postal services, and private households. Several other industries, such as employment in government hospitals and employment in government liquor stores, had to be added.

One of the most striking discoveries of the business list comparison project relates to the aggregate employment numbers. As shown in Table 2-1, the overall employment count in 2001 is 5.5 percent higher for the CBP than it is for the QCEW. These findings were reported in Foster et al. (2005), which also concluded that industry and geographic coverage matter a good deal and that the heterogeneity of results increases at finer levels of industry and geography. One purpose in cataloging the differences in published statistics was to help guide the micro-level analysis, which, given the 8 million establishment records, needed to focus first on the industries and states showing the greatest discrepancies.

Different methods were used to compare data from single-establishment units and multiestablishment units. Single-establishment businesses were examined on an exact match basis; that is, for a given establishment, the lists had to show the same number of employees to

[1]A full explanation of the methods used in this comparison project can be found in Foster et al. (2005).

count as a match. For multiestablishment businesses, the analysts used a "near match" band of 5 percent (± 2.5 percent). The percentage of cases in which the BLS and Census Bureau lists disagreed on single versus multiunit status was not available at the time of the October presentation.

In Table 2-2, the first row of figures shows the difference in aggregate employment for single-establishment businesses. The employment numbers are reasonably similar, but analysis of the microdata reveals important differences. About 30 percent of the employment and payroll estimates for the matched single-establishment businesses do not match exactly. Considering that the majority of Employer Identification Numbers (EINs) are for single-establishment businesses, these differences, both in number of establishments and the employment counts, are noteworthy and should be further explored.

In the case of multiestablishment businesses, using the near-match concept, the employment and payroll estimates match only 39 and 51 percent of the time, respectively. Again, this shows that significant micro-level heterogeneity underlies the comparatively similar macro-level statistics. The comparison project will document and further explore these similarities and micro-level disparities. In addition, the project will examine inconsistencies between the Census Bureau and BLS classifications of single versus multiestablishment businesses. As Table 2-2 indicates, there are approximately 309,000 cases (found by summing 197,000 and 112, 000, the bottom two rows of column two) in which the Census Bureau and BLS disagree over single versus multiunit status, and these businesses account for about 21-22 million employees. The project will examine a number of other topics—the role of nonemployers, the data quality for professional employer organizations and help-supply services, overlap and duplication in the COS and the MWR, and the role of firm identifiers—that factor into the inconsistencies found in the two business lists.

Ascertaining the sources of the nonmatched data will take time, as nonmatches are complicated with technical issues of scope and coverage and the cooperation of the states. Only 47 states authorized BLS to share their data for this project, and the relationship between the states that opted out and the nonmatches is still being explored.

Hanczaryk acknowledged that sharing between BLS and the Census Bureau would likely lead to improvements in both lists. BLS industry coding, physical location addresses, multiunit data from the MWR, and employment data for single units are recognized as being very thorough, and this detail would benefit the Census Bureau. The Census Bureau is particularly interested in the data for multiunit companies within states, as well as in BLS data for the client businesses of professional employer organizations (PEOs). PEOs (or employee leasing) firms typically supply human resource management services (e.g., payroll accounting or ben-

TABLE 2-2 Business List Comparison

2001 Microdata for 47 States	EIN (in millions)	BLS Employment (in millions)	Census Bureau Employment (in millions)	Based on Microdata Comparison	
				Match on Payroll	Match on Employment
Matched single-establishment EINs	4.1	35	38	71% (exact)	69% (exact)
Matched multiestablishment EINs	.112	49	48	51 (near[a])	39 (near[a])
Matched EINs, BLS multiestablishment, Census Bureau single-establishment	.197	15	15		
Matched EINs, BLS single-establishment, Census Bureau multiestablishment	.112	6	7		

NOTES: EIN = employer identification number, BLS = Bureau of Labor Statistics.
[a]Near match within ± 2.5%.
SOURCE: Workshop presentations by Jim Spletzer (BLS) and Paul Hanczaryk (Census Bureau).

efits administration) to their clients. The Census Bureau's tax record-based data do not accurately indicate the geographic location and industry of leased employees working at client sites; rather, they indicate the industry and location of the PEO itself.

BLS will benefit from an evaluation of firm information that is collected as part of the Census Bureau's COS. Access to the Census Bureau data could potentially add consistency to BLS industry codes, giving the agency the ability to analyze microdata on nonemployer businesses (18.6 million on which the Census Bureau publishes data).

Hanczaryk provided an overview of the current limited data sharing between the agencies. The Census Bureau provides BLS with approximately 1.2 million EINs every quarter, which BLS then matches to their files to provide industry codes and physical location addresses. From this process, for 2004, 3.4 million BLS industry codes were returned to the Census Bureau. Sharing these EINs and codes reduced costs and respondent burden and provided greater uniformity of the two agencies' economic data and, in the process, produced evidence that this type of data sharing works.

The comparison project currently under way will provide some indication to the agencies of what areas will provide the biggest payoff from more extensive sharing. Since, under the current agreement, the data can be shared for research purposes only—but not to update either of the registries—an important aspect of the project is to guide programmatic opportunities. The comparison would provide key input for any future reconciliation of data, should that become an option. For example, the COS and the MWR are now overlapping mail-out surveys to multiunit companies. By combining these two surveys, the agencies could reduce response burden on businesses, one of the twin goals outlined in CIPSEA.

In order to move forward on business list improvement and data sharing, the agencies must overcome important analytical and legal hurdles. First, the agencies need to better understand the magnitude of the differences and the reasons for them. However, comparing multiunit companies is complicated by the fact that there are no numerical identifiers that provide a one-to-one comparison of establishments. Second, without companion legislation, BLS is not authorized, under Title 26, to receive the Census Bureau microdata that include federal tax information. Finally, BLS, which has an explicit relationship with state unemployment insurance programs, would like to increase consistency in survey processes and economic data development, but this goal is hampered by the fact that the states cannot access key Census Bureau microdata.

Hanczaryk concluded that the potential of data sharing to improve business lists and other programs in the U.S. statistical system that would benefit users should provide BLS, the Census Bureau, and BEA with ample motivation to move forward.

AN OVERVIEW OF TAX DATA AND IRS DATA-SHARING ARRANGEMENTS

The availability of federal tax information, especially confidential microdata, was a recurring theme of the workshop. The IRS houses data on tax returns from a variety of entities, including individuals, estates and gifts, tax-exempt organizations, and businesses and corporations (tracked by EIN, not establishment). The IRS presenters—Mark Mazur and Nick Greenia from the Office of Research, Analysis, and Statistics—identified constraints to using data processed from these returns, most notably the need for an authorized purpose as defined by the Internal Revenue Code (IRC). Their paper (see Chapter 6) discusses the legislative history and lays out limitations, lessons learned, possible means to expand data sharing under current constraints, and steps that might be taken to change the legislation. Their presentation touched on three specific topics related to expanded access—the need for statutory change, regulatory change, and policy agreements; the importance of linking any expanded access to a specific research or statistical purpose; and the need to make the benefits to the Treasury Department and other agencies clear to policy makers.

There are three major authorized uses of federal tax information: tax administration (which is the concern that overrides all others from the IRS perspective), tax policy analysis, and research and statistical uses. Tax data are subject to strong presumptions of confidentiality, and Section 6103(j) of the statute authorizing access for outside statistical use permits recipients to access only the minimum amount of data necessary to accomplish a stated purpose. The penalties for unauthorized disclosure or inspection of tax data are strong and clearly defined in the statute. In addition, data recipients are subject to regular safeguard reviews of physical and computer security, need and use, and other factors in order to ensure that they are in compliance with IRS requirements to protect taxpayer confidentiality.

An increasingly difficult problem in releasing data is to ensure that they remain anonymous. Ever-advancing technology—improving data linkage programs, faster and cheaper computer processing, and increasing amounts of administrative data available on the Internet—has made it easier to match an individual or company to a specific record. For example, if a record includes an industry code, address, and revenue figure that could identify a particular business, anonymity would be breached and the law broken. In turn, protecting the confidentiality of tax data becomes increasingly more difficult.

The IRS considers all tax data sensitive, meaning that no distinction is made between information that is publicly available elsewhere, such as an address found in a phone book, and information that may not be, such as a firm's income and profit entries. In addition, no statute of limitations exists for tax data, which means that they must be protected in perpetuity. Various levels of security have been implemented to protect data from unauthorized use. The IRS maintains an audit trail for those who have access to the data, conducts background checks on users, and requires users of tax data to have a computer system that is separated from other types of data uses. These safeguards and constraints increase the cost of providing data and ultimately limit access.

Both IRS presenters emphasized the point that voluntary compliance is a cornerstone for accomplishing the central IRS mission of administering tax collection, and it is dependent on the protection, both real and perceived, of taxpayer confidentiality. Due to the sensitive nature of the information reported by taxpayers, ensuring and demonstrating the protection of confidentiality in the IRS process are vital aspects of promoting compliance. An ongoing concern at the IRS is that expansion of data access could increase the risk of a confidentiality breach and, along with it, a public perception that tax information is shared carelessly throughout the federal government. This in turn could weaken the voluntary self-assessment system. Mazur noted that a 1-percentage point change in the overall voluntary compliance rate translates into tens of billions of dollars in tax revenues collected. Thus, there are two important goals in managing tax data: protecting them to maintain as high a compliance rate as possible, and exploiting them effectively and efficiently for other authorized purposes, including statistical uses.

The characteristics of tax data are unique. The size of the population covered (over 20 million organizations and over 125 million individual taxpayers) and the scope of return data (covering information on everything from mortgage interest deductions to corporate net profits) create a complex respondent universe and a wealth of detailed data. Because of disincentives dissuading false or late reporting, nonresponse is thought to be low relative to most survey alternatives. However, the IRS captures neither all data for all types of returns nor data on taxpayers who fail to file; as such, there are well-known and systematic inaccuracies in the data reported to tax authorities. Nonetheless, given that the data are used for tax administration purposes, including enforcement and internal research and analysis at the IRS, there is reason to believe that many components of the data are accurate (again, relative to survey-sourced data).

The federal executive and legislative branches conduct tax policy analysis with the data, while four agencies covered in Section 6103(j) of the IRC—BEA, the Census Bureau, the Department of Agriculture's National Agricultural Statistics Service, and the Congressional Budget Office (CBO)—can use the data for statistical purposes, although the extent of access for specific purposes varies widely by agency. Data in nonidentifiable form, including public use files, have been used more broadly by decision makers in the federal government, businesses, policy think tanks, academic researchers, and state governments.

Statistical agencies are charged with using existing data systems, such as administrative records, to the maximum extent possible to reduce costs and counter concerns about respondent burden. The IRS, by contrast, must protect federal tax information by providing it only for authorized purposes and to the minimum extent required for each purpose. These two directives create tensions, some of which have been partially relieved by agreements that stipulate clearly delineated uses of tax data and the conditions under which they may be used. Generally speaking, however, the IRS does not view either burden or cost reduction alone as reason enough to grant access. Recognizing the great demand for access to federal tax information, Greenia and Mazur offered guidance about what kinds of sharing arrangements were most feasible.

As outlined in the Greenia-Mazur paper (see Chapter 6), three methods are available to expand access to data: statutory change, regulatory change, and policy agreement. Greenia clarified the differences between statutory and regulatory change: the former requires the passage of a law through Congress and the signature of the president, while the latter requires Treasury Department approval.

In order to add new statistical users, or to expand the access of those currently authorized to access federal tax information, Section 6103(j) must be amended. This statute stipulates who may use the data and for what purpose, as well as what data fields may be accessed—for example corporate tax items. The presenters suggested that limiting the specific data items to those actually needed for a specific purpose (e.g., basic information needed to construct a business list sampling frame), clearly tying requests to intended use, and specifying them in an amended statute might improve the chances of passing a proposed legislative change. The idea is to conform to the "minimum need" requirement by bounding the item content in the statute. Data sharing *among* authorized recipients is enabled elsewhere in statute—by Section 6103(p)(2) and the associated regulation (B).

On the regulatory side, the Treasury Department requires a compelling, data-driven business case in order to grant access to additional data items. Regulatory changes have been used in the past to both add and remove data field access. A change in regulation can supplement statutes and adjust for changes in user needs. For example, if the Census Bureau needed to access additional corporate tax data fields to accomplish a mandate under Title 13, Chapter 5, a change in regulation would be required to enable such an expansion in access. For purposes of business list comparison and reconciliation projects, Greenia discussed freezing item content in the statute to basic sampling frame information—again to be responsive to the minimum need requirement. He suggested, further, that limiting requests in this way would ease potential concerns by legislators and their staffs that additional items might be added in the future with only the approval of Treasury's assistant secretary of tax policy.

The third mechanism for expanding access is a policy agreement, which is intended to supplement the statutes and regulations. The Census Bureau-IRS Criteria Agreement, which emerged in response to concerns over access to tax data facilitated by the development of the Census Bureau Research Data Centers (RDCs), is an example. The crux of this agreement is that the work must have a predominant Title 13, Chapter 5, purpose, which essentially means it must improve Census Bureau programs. The Census Bureau is responsible for evaluating proposals based on scientific merit and predominant purpose, and the IRS documents the request and either authorizes or denies it after determining whether it complies with their regulations. External researchers apply for Special Sworn Status (SSS), go through an FBI clearance, swear to an oath to abide by Title 13, and then, after clearing these hurdles, are certified to access Title 13 and Title 26 data in the same manner as Census Bureau employees.

The Joint Committee on Taxation publishes an annual report that lists the volume of tax data records disclosed by the IRS, classified by statute and including statistical purposes. Disclosures under Section 6103(n), such as contract work with BEA or the Federal Reserve Board, are not included in this report. This report shows that the number of disclosures for federal statistical uses, most of which involve demographic data requested by the Census Bureau, is second only to those needed for state tax administration. Greenia noted that policy-oriented research and statistical analyses are important considerations for tax data administration. When asked about the effect that outside researcher access has on public perceptions of confidentiality and privacy, Mazur suggested that it is likely minimal, as long as the number of analysts is small, access is at arm's length, and researchers are subject to the same enforcement rules as others with data access.

During the open discussion, steering committee member John Haltiwanger inquired about possible models for new data sharing. One suggestion was to allow a third party, such as the Office of Tax Analysis in the Treasury Department, to access the data specifically for tax policy use and, second, to generate simulated synthetic data as public use data. Under this scenario, researchers from the Office of Tax Analysis would access data as SSS agents and follow the same rules as other external researchers working under the Census Bureau-IRS Criteria Agreement. The presenters recognized the potential importance of synthetic data as a tool for expanding access to microdata, citing several projects already under way at the Census Bureau. While the IRS is supportive of the Census Bureau work developing synthetic data, the quality and utility of the data sets are still largely unknown. Mazur noted that, while further development of synthetic data is necessary, past efforts have not generated data sets known to provide inference-valid results, especially for complex modeling applications.

In the context of moving forward on data sharing, the presenters offered a list of lessons learned (see Chapter 6 for a full discussion of these lessons). First, strong leadership and support from the highest levels of government are needed. The Office of Management and Budget (OMB), the Council of Economic Advisers, and congressional staff are good places to enlist support for a companion bill to CIPSEA. The effort needs active congressional support, from both staff of tax-writing committees and from members of the House and the Senate. It also requires clear communication to policy makers and the public of the potential benefits from interagency data sharing, such as increased efficiency, more accurate statistics, and reduced respondent burden.

Second, Mazur suggested that the myth of a zero-sum game, in which expanding access in one area requires reduced access elsewhere, must be addressed. In addition, he argued, discrete and incremental steps may be better than bold leaps as statutory or regulatory changes are pursued. Finally, there should be some interagency coordination of confidentiality protection procedures, and the benefits of proposed changes to the Treasury Department need to be clear. (In 2002, staffers from the Ways and Means Committee and the Senate Finance Committee asked what was in the companion bill for Treasury.) By leading the Treasury Department effort to advocate for CIPSEA in Congress, the IRS demonstrated that it can play a major role in the development and passage of data-sharing legislation. The more agencies that are behind the legislation, and the stronger the argument for widely distributed benefits, the more likely it is to receive congressional support.

DATA SHARING AND BEA PRODUCTION OF ECONOMIC STATISTICS

The extent to which agencies are able to share data for statistical purposes carries direct ramifications for national income accounting. As background for their presentation, Dennis Fixler and Steven Landefeld of the BEA contributed a paper on this topic, which appears as Chapter 7 of this volume.

In his presentation, Fixler specifically outlined how disparate sources of data can lead to inconsistencies in the construction of economic statistics. A prominent example is the fact that gross domestic income (GDI), gross domestic product (GDP), and state personal income have all, at times, displayed different rates of growth—differences that can present problems for policy makers. For example, GDP estimates are used for OMB and CBO budget forecasts, and the Federal Reserve incorporates the output growth information in monetary policy decisions. Chapter 7 provides details and further examples of why understanding these statistical discrepancies matters.

GDI and GDP are the two aggregate measures of domestic output. The product or expenditure side (GDP) is calculated primarily from the Census Bureau data, while the income side (GDI) is calculated mostly from BLS data. Conceptually, the two series should be equal, but because the data come from different sources, typically there is a statistical discrepancy. Historically, more weight has been given to GDP; however, as the statistical discrepancy increased during the latter half of the 1990s, analysts (e.g., those studying productivity trends or forecasting tax revenues) began paying more attention to the income side. Wage and salary growth for the period 1995-2001, as measured using the Census Bureau data, has been greater than that shown by BLS data.

Fixler and Landefeld suggest that one possible source of the GDP-GDI discrepancy may be tied to the way that stock options, bonuses, and fringe benefits are recorded in the BLS and Census Bureau payroll figures. Table 7-8 (see Chapter 7) breaks down differences in payroll growth to an annual basis, revealing that the Census Bureau figures are frequently but not always greater than those of BLS. To gain a clearer picture of measured wage trend differences, it is helpful to explore the data at the industry level. For example, between 1998 and 2002, the Census Bureau data show a faster growth rate than BLS data for the information sector; the opposite is true for construction. Calculating real value-added growth in a few selected sectors using Census Bureau rather than BLS data shows that the absolute differences can be substantial. The growth rate for computer and electronic products in 2002 illustrates this difference, as the Census Bureau measure of current-dollar value-added is roughly double the BLS count. The higher Census Bureau number supports an altered view of that sector, and of trends in manufacturing generally, suggesting a different recovery story for the period. More complete data sharing among the agencies would allow researchers to investigate these data discrepancies in a systematic manner.

Data sources used by BEA to compute gross output per worker—indicators of productivity—also show substantial differences. Two examples, shown in Table 2-3, are oil and gas extraction and petroleum and coal products, in which the percentage differences are 13.9 and 12.8, respectively, between the CBP and QCEW measures of gross output per worker. These differences are linked to those discussed in the presentation by Spletzer and Hanczaryk on the BLS and Census Bureau business

TABLE 2-3 Gross Output per Worker (in dollars)

2002 NAICS Code	Selected Industries	2002 Output per Employee		
		Census Bureau	BLS	Percent Difference
211	Oil and gas extraction	991,595	853,547	–13.9
324	Petroleum and coal products	2,062,617	1,798,598	–12.8
486	Pipeline transportation	761,076	660,673	–13.2
515-517	Broadcasting and telecommunications	296,694	342,739	15.5
52-535	Finance, insurance, real estate, rental, and leasing	392,955	434,753	10.6

NAICS: North American Industry Classification System.

list comparisons, and they show exactly why BEA has an interest in this project.

Another area for further exploration relates to Title 26 data. BEA, the National Science Foundation, and the Census Bureau worked together on a project begun in 2003 looking at research and development expenditures. The project was designed to use Title 13 and BEA data, avoiding federal tax information. As Table 7-10 reveals (see Chapter 7), in some cases the expenditures of U.S. parent companies, a subset of all U.S. firms, exceeded all U.S. expenditures. This inconsistency between expenditure measures could potentially be solved by using Title 26 data. Estimation of state income taxes provides another example of how multiple sources of data can lead to inconsistencies in accounting. Chapter 7 includes estimates of the extent to which the difference between the BLS and Census Bureau payroll figures can affect projected state and local income taxes.

The benefits of data sharing can be viewed from either a system-wide perspective or, more narrowly, from the perspective of specific agencies. The system-wide benefits include improved sampling frames, more consistent industry and region classifications, and an increased capacity to resolve anomalies in responses—all without increased respondent burden. Finally, there are important analytical policy questions that can be addressed at the micro level through data matching. Prime examples are policies involving foreign direct investment and offshoring.

From the national accounts perspective, Fixler argued that data sharing offers a number of benefits. Data sharing would aid in resolving the statistical discrepancies underlying source data, such as those underlying the income and product sides of the national accounts and the payroll and employment estimates. Preliminary employment figures, indicators, and extrapolators, which are used in BEA's early estimates, as well as its models and projections, may be improved through data sharing and fuller access to data that capture accounting and other business changes. The extent to which sharing would help overcome problems of data disruptions, such as from natural disasters like Hurricane Katrina, is unclear, but allowing agencies to compare notes could help fill the gaps.

During open discussion, Carol Corrado noted that some discrepancies are informative. She pointed out that analysts at the Federal Reserve Board look at differences across data series to piece together analytical insights. Identifying sources of discrepancies requires access to data to determine if there is a difference in exporting between the Census Bureau and the BLS forms, a difference in the interpretation of language, or something else. Fixler noted that, under current constraints, this capability is limited, as restrictions on federal tax information typically do not allow for adequate analysis below the aggregate levels. Fixler stressed that data sharing cannot solve all discrepancies, but it will allow analysts to better understand the source of the differences and provide policy makers with a clearer picture of what is happening in the economy.

3

Research and Policy Perspectives on the Benefits of Business Data Sharing

Business data from the U.S. statistical system and various private-sector sources form the country's economic informational infrastructure, on which millions of policy and business decisions, including cost-of-living adjustments, monetary policy, and wage and investing decisions, are based. Shortcomings or inconsistencies in measures of economic activity can hinder decision making in a multitude of ways. Steering committee chair Charles Schultze introduced the afternoon session by taking up this topic, noting that a lack of data sharing and coordination can contribute to problems of discrepant data. One notable example is the aggregate employment statistics, specifically, the (perhaps cyclical) pattern of differences in employment changes as measured by the Current Employment Statistics (CES) and the Current Population Survey (CPS). Schultze cited the widely held view that, in the short to medium run, the CES payroll series is the best measure for judging cyclical strength or weakness of the economy, but he suggested that policy makers and analysts should pay attention to the employment estimates in the CPS household survey as well and support efforts to reconcile the two series.

The discrepancy in employment data has recently received a good deal of attention. Schultze cited a paper by George Perry (2005) evaluating the CES and the CPS (adjusted for conceptual comparability) in terms of their usefulness for measuring the short-run employment behavior of the overall economy. He also cited the work of John Schmitt and Dean Baker (2006), economists at the Center for Economic and Policy Research, which found that the CPS may be overstating the share of working Americans by 1.4 percentage points (roughly 3 million people).

The choice of how much reliance to place on either of the two employment series affects measured growth of unit labor costs and productivity, which carries important policy implications. For example, the higher unit labor costs derived from the CPS (relative to the CES) would support a stronger case for increasing interest rates. Schultze agreed with the position that both data series contribute relevant information, but he added that the reasons underlying the inconsistencies need to be more fully understood. If it were possible to integrate Census Bureau data on the self-employed and nonemployers with employment statistics from the Bureau of Labor Statistics (BLS), one element of the discrepancy in establishment- and household-based employment statistics could be studied more carefully.

The afternoon session featured the perspectives of data users outside the statistical agencies. Carol Corrado opened the discussion of user needs and concerns related to data sharing with a presentation on monetary policy and research at the Federal Reserve Board (FRB). Dale Jorgenson and George Plesko provided additional researcher perspectives with presentations on productivity and real output measurement and on tax data needs for estimating corporate profits, respectively. The session concluded with a presentation on the use of data and the potential value of data sharing for budget forecasting by Douglas Holtz-Eakin, former director of the Congressional Budget Office (CBO).

MONETARY POLICY AND RESEARCH AT THE FEDERAL RESERVE

Carol Corrado spoke from the perspective of the FRB—perhaps the most important policy consumer of economic statistics—about the value of accurate data and how improved coordination among the statistical agencies might enhance that value. Her comments focused on how business list reconciliation would provide a more consistent source of data for use in the analysis and forecasting of productivity—particularly multifactor productivity. Real-time economic statistics, imperfect by nature, play an important role in the Federal Reserve's assessment of underlying trends in gross domestic product (GDP) and gross domestic income (GDI), payroll and household-based estimates of employment, industry productivity, and business inventories. Corrado cited these and several other examples as cases in which alternative data sources do not typically align. Sometimes one source is more revealing than another; for example, data from the income side of the national accounts captured the 1990s acceleration in productivity considerably sooner than did data from the product side.

In some cases, alternative measurement instruments reveal different elements of a phenomenon, thereby allowing richer analyses; in other cases, they do not. Payroll and household employment statistics, which frequently diverge, are useful in revealing different aspects of the labor picture. However, Corrado argued, maintaining two business establishment lists is not a good example of complementary data sources. She emphasized the point that there is a distinction between data consistency and data integration and the way each relates to statistical discrepancies. Consistency has to do with definitions and classification, which is what the reconciliation of the two lists should aim to achieve. She suggested that the methods by which the Bureau of Economic Analysis (BEA) reconciles its industry accounts with the expenditure side of GDP is an example of integration eliminating informative discrepancies. Fixler noted that, from a national income accounting perspective, BEA analysts need to be able to draw as much information from one account toward the other as possible. If the differences were fully understood between the inputs, the decision to integrate or distribute discrepancies between accounts would be unnecessary, as the differences would be small and relatively manageable. Corrado suggested there may be problems with the way discrepancies get redistributed in the national income and product accounts, and that noise in the data needs to be eliminated before they can be used to detect changes in productivity at the industry level.

Corrado used the striking productivity measurement case to illustrate problems that arise from maintaining two business lists. At the FRB, aggregate productivity is directly linked to monetary policy decisions because productivity affects inflation trends. Conventional industry-based productivity studies rely on output data from the Census Bureau and input measures from the BLS. Corrado pointed out that a calculation in which the numerator and the denominator are derived from the same survey or survey frame would have obvious statistical advantages.

One example of the importance of industry-level information for assessing prospects for aggregate productivity surfaced during the late 1990s. Initially, in late 1996, the FRB's view of productivity was based on analyses of sectoral and industry trends captured using BEA's data on GDP by industry. These analyses, along with other academic and agency research, suggested that prices in some industries were mismeasured. As the role of changes occurring in the high-technology industries became clear, a strong case emerged for improving industry-level data. Still later, evidence grew that the step-up in multifactor productivity was fairly broadly based by industry. Although these developments are now apparent in retrospect, at the time these new trends were emerging, when the FRB needed to examine what was happening in productivity in real time, much of the available evidence of the changes in multifactor productivity was anecdotal.

When measuring productivity, most macroeconomists do not spend a lot of time with the Census Bureau employment data, mainly because those figures are not comprehensive (even in Census years). Macroeconomists more typically rely on BLS industry-level employment measures, which at times diverge significantly from Census Bureau-based measures. For example, in 1997, 2.6 million persons were categorized as working in the management of companies (North American Industry Classification System, or NAICS, Code 55) according to the Census Bureau, but only 1.7 million persons worked in the occupation according to BLS. A reconciliation of the Census Bureau and BLS business lists has the potential to create more uniform and comparable measures of output, employment, and, in turn, productivity. Although anecdotal information will always be used to some extent, Corrado argued that a system that includes a consistent historical time series is vital to research and policy at the FRB.

Corrado commented on additional data hurdles that inhibit analysis at the Federal Reserve, citing lags in data, the impact of the switch to NAICS on productivity analysis (particularly for the services industries), and BEA changes in its industry accounting methods. She noted that the FRB is engaged in a project to review disaggregate productivity measurement using industry data from BEA, as well as aggregate measures for six sectors believed to illuminate key trends. Inconsistencies in employment by NAICS-defined industries have limited what the project is able to do in terms of studying trends in multifactor productivity. Project researchers built a Standard Industrial Classification (SIC)-to-NAICS concordance in an effort to bridge some gaps and, in Corrado's view, fully reconciling the business lists would advance the goal of creating time series of consistent historical industry productivity data. Once current employment data discrepancies are addressed, she would like to see a historical revision of industry output and productivity data.

Corrado concluded her presentation by stating that the statistical agencies should reorient their programs to avoid duplicative effort required to maintain two systems, which ultimately makes detecting trends in the economy more difficult. During the open discussion, John Haltiwanger noted that it would be difficult to choose (or to get BLS and the Census Bureau to agree) on one list over another, and he urged that the strengths and weaknesses of each would need to be captured to construct an ideal downstream reconciliation of the two. Historical estimates could continue to rely on the Census Bureau employment data because they conform to the Census Bureau output data but, looking ahead, the list would benefit from the timeliness of BLS data. Agreeing that this would be an appropriate strategy, Corrado expressed the hope that, in the near future, there would be a system going forward that has reconciled the differences in employment by industry.

PRODUCTIVITY AND REAL OUTPUT MEASUREMENT

Dale Jorgenson presented his views on the role of data sharing in the measurement of productivity and real output and on data integration issues related to the "new architecture" he helped design for the U.S. national accounts. A goal of the new architecture is to provide an internally consistent and comprehensive set of national accounts to measure the large, complex U.S. economy with a statistical system that is highly decentralized. He argued that an accurate set of accounts depends on a consistent set of data. One goal of the new architecture program is to have a unified register of firms and establishments, collecting data in a way that reflects the common records that are maintained for these units; some of this can be accomplished through more effective data sharing by the statistical agencies. Such a system would work to achieve greater internal consistency of data at the micro level, helping to identify the sources of discrepancies.

The new architecture framework has seven main accounts: production, domestic receipts and expenditures, foreign transaction current account, domestic capital account, foreign transaction capital account, domestic balance sheet, and U.S. international position (see Figure 3-1 for

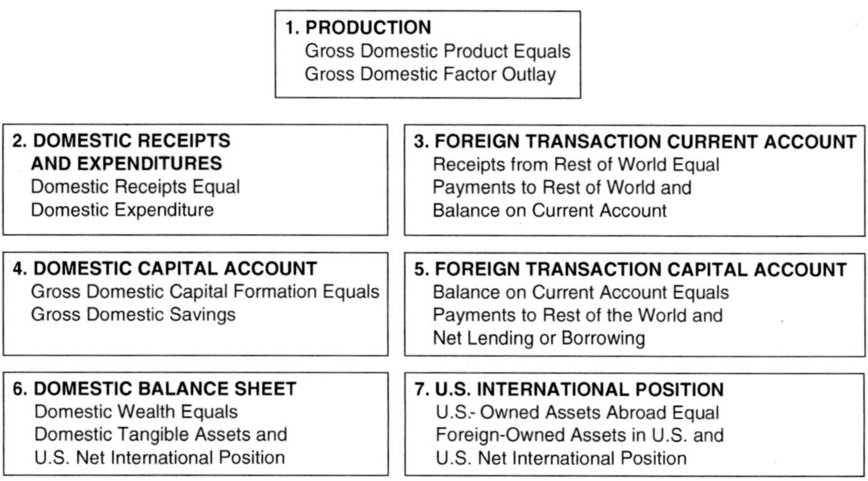

FIGURE 3-1 Blueprint for an expanded and integrated set of accounts for the United States.

detail on components of each account). The majority of Jorgenson's presentation focused on the production account, which sets GDP equal to Gross Domestic Factor Outlay or Income (GDI).

The production account consists of output, input, and productivity, in which output and input shares add up to one (1.0) in the GDP. Consumption and investment outputs make up the output shares, and when one rises, the other falls. GDI reveals how individuals divide their money among resources for consumption (which includes immediate consumption goods) and savings (which includes housing and financial assets). GDP and GDI should agree, but in order for those to match, the source data must be internally consistent. When these are consistent, they can be used to measure the effects of events like Hurricane Katrina on the economy. However, when the components of the accounts do not agree, errors within the accounts and among the seven accounts occur. For example, consumption expenditures show up as part of the national product. An error in the product account leads to an error in expenditures and income, which in turn creates an error in production. In addition, discrepancies in the measurement of the savings rate cause problems in the capital formation and wealth area. Thus, error propagates error.

BEA is responsible for the production accounts at both aggregate and industry levels. Input shares to productivity include capital and labor inputs. The estimates of labor inputs are based on BLS data from the Quarterly Census of Employment and Wages and have remained fairly constant. If the Census Bureau figures were used rather than BLS numbers, the result would be a share of labor about 5 percent lower and the share of capital would increase. This would affect the estimate of economic growth on the input side and, in turn, the productivity number. In terms of contributions to growth, capital investments are primary sources of economic growth, because labor growth has been slow relative to capital and investment growth since 1995 (see Figure 3-2). If adjusted to Census Bureau data, where the labor share is even smaller and the capital share larger, over time productivity estimates would decrease. Jorgenson provided these examples to illustrate how data choice (in this case, the Census Bureau versus BLS data on shares of labor and growth measures) affects the internal consistency of the accounts and the ability to obtain accurate macroeconomic measures.

Additionally, the output data produced from different data sources by BLS and BEA are not comparable, and it is important to be able to reconcile these differences. One way to do that is to go to the registers of individual firms and establishments and make sure the data are collected from common sources. Jorgenson reiterated that accurate measurement of output, productivity, and saving requires extending data-sharing arrangements. He argued that it is necessary to work toward a unified reg-

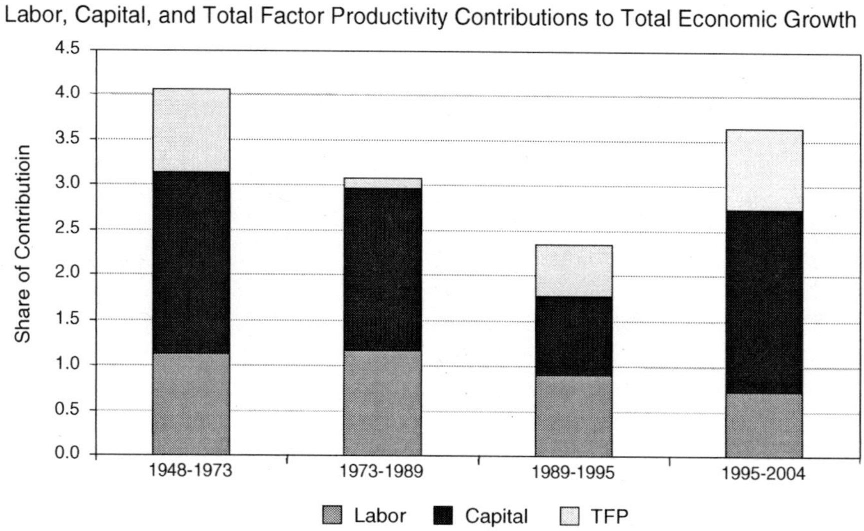

FIGURE 3-2 The production account—output, input, and productivity.

ister so that the national accounts can rest on a firm foundation of a common set of data sources. He added that having a centralized statistical system, such as Statistics Canada, is not necessary, if the problem of internal consistency is addressed. In order to have a common register that is used across the government and to share data among agencies, the statutory authority to do so must exist.

Jorgenson stated that data sharing is not an academic problem or an issue for the statistical agencies alone. It is a problem for the policy makers and the politicians who rely on the agencies for information. He speculated that policy makers are hamstrung by the fact that the data system involves inconsistencies that arise through the lack of data sharing and through the absence of statutory authority to share data among the statistical agencies. Jorgenson stated that it is absolutely essential that these uncertainties be resolved, and he suggested starting at the most fundamental levels.

For monetary and economic policy purposes, understanding the current state and future outlook for the American economy depends on getting the numbers right. One of the most important factors in determining the economic outlook is the contribution of information technology and other industries to past—and future—growth and productivity. To illustrate, Jorgenson calculated differences in productivity growth implied by the Census Bureau and BLS data. Differences at the industry level are typically much higher than the 5 percent that he found in the aggregate estimate. For manufacturing as a whole, the share of labor calculated using the Census Bureau data is 15 percent lower than the statistic calculated using BLS data. For some industries, such as oil and gas extraction, the Census Bureau figure is 50 percent lower than the BLS figure.

At the industry level, one of the most dramatic differences between value-added growth, as estimated using the Census Bureau data rather than BLS data on wages and salaries, is for the computer and office equipment industry. Estimates provided by Landefeld and Fixler indicate significant differences in the growth rate and level of activity in this industry and the broader information technology industry, depending on whether the BLS or Census Bureau data are used (see Table 7-2 in Chapter 7).

During the open discussion, Haltiwanger made the point that there seems to be a clear distinction between informative discrepancies and inherent, nonmeaningful inconsistencies that are costly in terms of time and resources. He suggested that the sectors of the economy experiencing large amounts of change might also be the most problematic. In considering whether the agencies are performing their statutory duty to provide the most accurate statistics possible, Jorgenson's two biggest concerns are with measuring output at the industry level and measuring the income side. He concluded with a clarification of an important discrepancy: if the macro number and the number calculated by adding up the industries do not match, that is an important indicator of discrepancies in the industry-level data across sources. In order to understand the future of the economy, he argued, the macro picture will not suffice, and the industry data from the FRB, BLS, and BEA are vital. Finally, he noted that, while not all of the issues can be solved by data sharing, such arrangements are essential to improving economic data and statistics.

TAX DATA NEEDS FOR IMPROVING ESTIMATES OF CORPORATE PROFITS

George Plesko of the University of Connecticut discussed the availability and role of various kinds of corporate profit information for research and analysis. He focused on discrepancies and measurement differences between financial reporting profits and cash reported profits, as well as how each relates to the other. His paper (Chapter 8) discusses how tax return data can be used to improve estimates of corporate profits. Corporate income can be measured in two ways, through financial (more commonly called "book") reporting and through tax reporting. Book reporting provides income measures and other information for investors, creditors, and other users; however, such information is available only for publicly traded companies. Tax reporting provides a measure of income for tax collection, and it is generally characterized as providing less discretion to management regarding the choice and application of accounting rules. Fuller data sharing might allow researchers to examine the relationship between tax and book data and the viability of using the latter in tax-estimate modeling.

Tax-based and book-based calculations of corporate income are different for two primary reasons—one causing a temporary difference and one a permanent difference. Tax depreciation is typically faster than book depreciation, leading to less income in the short term, but more in later periods—thus creating a temporary difference in the measurement concepts. Second, tax-exempt interest is not included as part of taxable income but is considered income for financial reporting purposes, which creates a permanent difference in measurement. These differences are important to outside investors, creditors, and policy analysts for understanding the operations of a firm. Recent research (U.S. Department of the Treasury, 1999) has also identified large and increasing differences in book versus tax income as supporting evidence of increased tax sheltering activities (see Table 8-1). The book and tax reports attempt to measure the same thing but the methods are conceptually different, and each adds something to what outside analysts can learn about a firm.

Financial statements provide an opportunity for analysts to augment information about a firm's operation that is not required to be disclosed through tax reporting. The information from each is useful to investors, and ideally these two sources would be complementary. However, the information is not useful for inferring income and payments. When examining pretax book income and the amount of income subject to tax from the two sources, Plesko found that, over time, the two diverge. In terms of nonpublicly traded firms, tax return information may be the only source of financial information.

The timeliness of book reporting is important to note, as information provided through this type of financial reporting is filed publicly by March 30 for the preceding year. Tax return information from Statistics of Income (SOI) tabulations lags behind by nearly three years when taking into consideration extensions and audits. Given the earlier access to book statements, if the relationship between a firm's tax returns and the book financial statements is clear, then the latter should enable analysts to begin modeling a year or two ahead of the release of SOI figures. A problem arising from using financial statements is that more than one, and possibly several hundred, matching tax returns may exist. The new IRS Schedule M-3, which replaced Schedule M-1, allows fuller reconciliation of book income with tax net income. The form delineates differences between book and tax income as temporary or permanent, allowing every tax return to be identified with the appropriate accounting parent.

When combined, financial and tax return information provides a more complete picture of a business's operations. Access to multiple data sources also allows for more timely estimation of economic events because various sources become available on different schedules. The ablity to reconcile book and tax income at the back end creates benefits as well. For example, financial statement data are useful in augmenting tax filings by providing information on firms' operations that are not captured in the tax reporting system.

The new IRS schedule, the M-3 (see http://www.irs.gov/pub/irs-utl/2005f1120sm3.pdf), is coming online beginning with tax year 2004 for firms that have assets of $10 million and more. Benefits of the Schedule M-3 include reconciliation of income from the worldwide consolidated financial statement and from the income statement of includible corporations; reconciliation of book income of includible corporations with tax net income; and delineation of differences between book and tax income as temporary or permanent. The new form should help the IRS, Treasury, and BEA better understand how financial reporting can inform tax administration.

Plesko cited the national accounts as an example of how the new data that incorporates both financial and tax return information will be useful. The detail captured by the M-3 will provide more data on the contemporaneous finances and the operations and organization of firms, including specific decisions related to tax planning. Plesko argued that the improved financial data that can be generated from the M-3 will be helpful in measuring the effects of tax policy, as well as for constructing preliminary national income estimates of corporate profit (see Chapter 8 for further examples). During the open discussion, steering group member Daniel Feenberg suggested that BEA and the Census Bureau think about cooperating on a reconciled form that would allow BEA to ask respondents to reconcile their answers with those provided to the Census Bureau.

THE IMPORTANCE OF ACCURATE DATA AND DATA SHARING FOR BUDGET FORECASTING

The CBO is charged with providing Congress with the information it needs to address budgetary and economic policies. CBO produces approximately 900 cost estimates per year, on tight time schedules and for difficult and disparate questions. CBO gathers data from agencies, including administrative data, and it relies on confidentially supplied private-sector data. It is often difficult to reconcile the different pictures of a program's performance portrayed by administrative data from the agencies and by data based on the Current Population Survey or other surveys. CBO has indirectly experimented with data sharing in the sense that it has made attempts to reconcile confidential private-sector data with data from other agency sources.

CBO uses data in a number of ways: for monitoring the state of the economy at both the macro and micro levels; for near-term forecasting (18 to 24 months) of overall movement and composition of output and income; and for medium-term forecasting for the economy over the next 10 years. CBO also engages in long-term modeling to produce estimates of the future path of various programs, such as Social Security and Medicare, analyzing changes in program finances and their impact on the federal unified budget deficit and the path of the U.S. economy.

In his presentation, Douglas Holtz-Eakin began by asserting that, of course, CBO would be better served if it had more accurate and timely data, and he suggested that an important consideration in evaluating the strength and accuracy of data is the underlying expertise in its production. To the extent that it is possible to enlarge the effective sample sizes through sharing of existing resources, budget analysts might be able to anticipate and answer questions before they arise. Holtz-Eakin suggested that timeliness of input data improves budget forecasting in an important way: when budget forecasts can be based on timely data, fewer revisions are required, leading to more consistent histories.

Revisions of historical information impede forecasting efforts. To the extent that revisions can be minimized through better accumulation of data, Holtz-Eakin suggested that the entire process will be improved, because attention will focus on the judgments that were actually made in doing the forecasts rather than the accuracy of the information going into them. If the data are revised less often, CBO and others will have more confidence in the forecast estimates they pass on to Congress.

The question for the workshop concerns the extent to which data integration permits better forecasting. Of course, there is a trade-off at some level between getting things quickly and getting them reliably. The issue is how much CBO can move away from human intelligence and heavy dependence on anecdotal evidence, particularly in the policy world, to data that are more reliably gathered. Landefeld noted that BEA strives to produce better early estimates that hold up over time. The question then shifts from more timely data to producing more accurate early estimates in order to make the trade-off between timeliness and accuracy less stark.

Holtz-Eakin noted that, recently, federal fiscal policy has been reactive. If better data were provided for monitoring the economy in real time, some activities in the federal government would see tremendous benefits. This might allow analysts to sharpen near- and medium-term economic forecasts. For example, real-time economic data would help to better elucidate shifts in productivity, inflation, core consumer prices, income and receipts, corporate profitability, bonuses and options, the composition of employee compensation, and labor force participation. Simply knowing, on a timely basis, when money comes into the Treasury—and whether it is income tax or payroll tax—would allow for better understanding of wage and salary movements.

Holtz-Eakin noted the importance of providing access to data used by CBO in order to improve the transparency of the processes, since the ability to explain how forecasting is done is just as important as doing it. For example, allowing outside analysts to produce a parallel set of estimates facilitates two important kinds of comparisons: comparisons that are "prickly and uncomfortable" but that provide information from the mistakes, and ones that provide transparency to the policy process. More important than the numbers themselves is the need to inspire confidence that they are constructed fairly, distributed in a timely fashion, and used in a manner that allows the policy process to evolve smoothly.

In Holtz-Eakin's view, it is not enough to give the people making economic and budget estimates carte blanche access. No one would know how to assess the relative merit of competing estimates unless the process of constructing those estimates is transparent. He suggested the need to push hard for the private sector to have access to data, not only for their own use, but also as a way to support work of the government.

In response to a comment that extending data sharing is not politically popular, a participant asked if members of Congress had an interest in this topic. Holtz-Eakin responded that a number of members of Congress do have an understanding of the sources of difficulty in making budget projections, and that they, in fact, were sympathetic. He added that there was political interest in improving data but conceded that confidentiality is a major concern. He suggested that the argument for access should be framed so that the benefits (such as improving the ability to accurately decompose sources of economic activity) are clear and expressed in a practical way.

4

Key Points from the Presentations: Directions for the Future

The workshop concluded with a summary presentation by Robert Parker, former chief statistician at the Government Accountability Office and the Bureau of Economic Analysis (BEA), in which he highlighted some of the key points made during the workshop. He also posed a few questions and offered suggestions to the agencies for going forward. Kathleen Utgoff, commissioner of the Bureau of Labor Statistics (BLS), Howard Hogan, associate director for demographic programs at the Census Bureau, and Rosemary Marcuss, deputy director of BEA, provided concluding remarks.

In a paper written for this workshop (see Chapter 5), Parker summarized the history of data-sharing legislation and various reports on the topic. Among his points were the following (the first three are from his paper and the last two from his presentation):

1. Prior to the 2002 Confidential Information Protection and Statistical Efficiency Act (CIPSEA), there was a significant amount of interagency business data sharing. Most Census Bureau and BEA access to tax return data was granted under Internal Revenue Code Title 26, Section 6103; more modest data sharing among BEA, the Census Bureau, and BLS was authorized by the International Investment and Trade in Services Act of 1990. For data sharing related to use in the economic censuses, the Internal Revenue Service (IRS) facilitated the success of this program by adding questions to selected tax returns solely for the purpose of the censuses.

2. Since the passage of CIPSEA, additional data-sharing programs have been undertaken and, based on the agency presentations at the workshop, others are in the planning stages.

3. The administration has not actually submitted its version of the "companion legislation." According to agency representatives, the Joint Committee on Taxation has attempted to develop its own version, but it has not met agency needs (there also seem to be different opinions as to whether the Treasury Department supports the legislation).

4. To expand the use of data sharing in the absence of passage of companion legislation, there should be a comprehensive review of IRS Publication 1075, which sets the rules that determine the extent of data sharing under Section 6103 by the IRS and the CIPSEA-designated agencies. This review could result in changes in the rules that might further facilitate data sharing within the provisions of Section 6103 (current rules were not developed in conjunction with the agencies, and there is no public record of the underlying legal interpretations).

5. Discussions have taken place between the Census Bureau and BLS about developing a single business register for *employer* firms (for which both agencies have data), perhaps in time for the 2007 economic censuses. Although the current business comparison project will yield a substantial amount of information about current differences, BLS and the Census Bureau presented no plans on how this information might be used to develop a unified business register (because of problems with various state laws, it may be necessary for BLS to maintain a separate national register as the sample frame for its establishment surveys).

Following up on the fourth point, Parker suggested that IRS and the statistical agencies could work to clarify the rules governing what the Census Bureau is permitted to do with data, such as those indicating business names and addresses, which, at the time they are received, are considered federal tax information. The Census Bureau must reenter the information on another form and send it to businesses for verification to become sharable data under CIPSEA. Because the IRS has never published legal interpretations explaining how they went from Section 6103 to these regulations, Parker suggested that the agencies and the IRS might collaborate to see whether or not current regulations are appropriate or necessary and provide input on what can be done to make it easier to convert tax information into Title 13 information.

Parker also reflected on Carol Corrado's presentation, particularly her point that maintaining two different establishment lists is unnecessary and counterproductive. Parker noted that the comparison project has a seemingly infinite research agenda, but that nothing concrete was presented by the BLS and Census Bureau presenters about producing a single list. He argued that users would ultimately gain little from that particular data-sharing opportunity unless there were plans for some implementation of a single list.

Kathleen Utgoff, Howard Hogan, and Rosemary Marcuss responded to Parker's remarks as well as to the workshop presentations in general. Utgoff and Hogan both discussed the various data-sharing projects and agreements that are planned or already taking place and the players in data sharing outside the IRS. They specified the various programs that would benefit from data sharing, reiterating the advantages of a single business list, and discussing the short- and long-term effects of CIPSEA on resources and on the work of the agencies. Marcuss suggested that the agencies seek advice on best practices from important data users and work toward increased access in the long term.

Utgoff responded to the call for a single business establishment list, acknowledging the potential value, but also noting that this is just one reason for extending data sharing. She discussed other aspects of BLS work that may be enhanced by data sharing, including improved unemployment and income measurement and development of better survey sample frames. She also noted the potential of data sharing to improve the capacity of state governments to forecast revenues and for underpinning economic development plans.

Utgoff reported that both BEA and BLS are motivated to improve data sharing and highlighted other data-sharing projects among BLS, BEA, and the Census Bureau. Two such agreements exist between BLS and BEA—one to share consumer price index data and another to share company and revenue data (collected under the International Investment Trade and Services Survey Act). Sharing the price index data will allow BEA to determine the feasibility of developing interarea price indexes, which could be used to adjust personal income estimates for price differences across geographic areas. New price indexes to measure royalties and license fees in international transactions could also be created from revenue data shared by BEA. The agreements between BLS and the Census Bureau authorize sharing of the BLS Quarterly Census of Employment and Wages data for use by the Census Bureau Longitudinal Employer-Household Dynamics program, as well as use of the Census Bureau Business Register by the BLS producer price index program to facilitate more efficient sampling.

In order to fully realize the benefits of CIPSEA, these interagency projects require a dedicated commitment of resources and support from the data-user and policy-making communities. One aspect of the overall benefit of these projects is the opportunity to reduce costs and respondent burden. Workshop participants paid particular attention to the data-sharing component of the CIPSEA legislation, but Utgoff acknowledged the importance of Subpart A, establishing uniform statutory confidentiality protections, which BLS has long needed as part of their effort to maintain survey response rates.

Hogan began his remarks with the point that conducting the research and development on data sharing marked the first real use of CIPSEA, as people learned about such things as the documentation required, the approvals needed, and the mechanics of applying it. Beyond IRS data, Hogan noted opportunities for data sharing in diverse areas, such as product line and price index information.

On the business list topic, Hogan stated that short-term goals can be accomplished while proceeding with long-range plans. For example, research can begin on the differences in the lists at the aggregate level. The Census Bureau has already seen benefits from list comparison work—for example, 10 percent more firms have been added to the research and development sampling frame, and costs and respondent burden have been reduced. Before a single list is possible, Hogan observed, tax data sharing and the role of the states (which provide and share data with BLS but not with the Census Bureau) must be addressed. Overall, there are a number of benefits to the Census Bureau made possible by data sharing, and he expressed excitement on behalf of the agency over the cooperative opportunities.

Adding to the comments of Fixler and Landefeld, Marcuss reiterated the BEA goal of bringing analytical energy to the task of demystifying data inconsistencies and making data from diverse sources fit. BEA will continue to adjust its methods and pursue new approaches to more fully exploit the currently sharable and available data. In addition, BEA seeks advice from the agencies and the IRS on the best, least intrusive, procedural changes in data-sharing practices that can improve estimates of gross domestic product. She added that the agency would quickly utilize information on the sources of differences between the Census Bureau and BLS data from any comparison work—for example, from a partial comparison of the business registers, even short of movement to a single list.

Marcuss added that advice from important data users—particularly policy makers in the administration and Congress, and businesses—on where the greatest weaknesses exist would continue to serve as valuable input. As a long-term goal, she hoped for increased data access for statistical purposes, confirming that BEA has already taken steps forward and is ready to participate in and use the information gained from the data-sharing projects. Her comments echoed those of the workshop in general. Overall, the agency representatives, workshop presenters, and other participants were supportive of extending data sharing to the extent that such arrangements can be safely implemented to increase the efficiency, accuracy, and consistency of the nation's economic statistics.

References

Foster, L., J. Elvery, R. Becker, C. Krizan, S. Nguyen, and D. Talan

2005 *A Comparison for the Business Registers Used by the Bureau of Labor Statistics and the U.S. Census Bureau.* Paper presented at the Joint Statistical Meetings, August 7, Minneapolis, MN.

Jorgenson, D., S. Landefeld, and W. Nordhaus

2006 *A New Architecture for the U.S. National Accounts.* Chicago: University of Chicago Press.

Norwood, J.L.

1995 *Organizing to Count: Change in the Federal Statistical System.* Washington, DC: Urban Institute Press.

Office of Management and Budget

2004 *Statistical Programs of the United States Government, Fiscal Year 2004.* Available: http://www.whitehouse.gov/omb/inforeg/04statprog.pdf [accessed May 2006].

Perry, G.L.

2005 Gauging employment: Is the professional wisdom wrong? *Brookings Papers on Economic Activity* 2:285-321.

Schmitt, J., and D. Baker

2006 *Missing Inaction: Evidence of Undercounting of Non-Workers in the Current Population Survey.* Washington, DC: Center for Economic and Policy Research.

U.S. Census Bureau

2002 *North American Industry Classification System—United States 2002.* Washington, DC: U.S. Department of Commerce.

U.S. Congress

2002 *Confidential Information Protection and Statistical Efficiency Act of 2002.* (HR 4528). Available: http://www.eia.doe.gov/oss/CIPSEA.pdf [accessed May 2006].

U.S. Department of the Treasury

1999 *The Problem of Corporate Tax Shelters. Discussion, Analysis, and Legislative Proposals.* (White paper, July.) Washington, DC: U.S. Department of the Treasury.

Appendix A

Workshop Agenda and Attendees

Workshop on the Benefits of Interagency Business Data Sharing
October 21, 2005
The Keck Center of the National Academies
Room 100
500 Fifth Street, NW
Washington, DC

Welcome and Introductory Comments

9:00 am Background, Workshop Goals, and Agenda
Charles Schultze, workshop chair *(Brookings Institution)*

9:10 Welcome from the Sponsor
Steven Landefeld, Dennis Fixler (BEA)

Benefits of Data Sharing to the Statistical Agencies

9:15 Improving Business Lists: The Business Register Comparison and Reconciliation Projects
Jim Spletzer (BLS); Paul Hanczaryk (Census Bureau)

9:45 Open Discussion

10:15 An Overview of IRS Data: Data-Sharing Arrangements, and Data Needs of Primary Customers and Researchers
Mark Mazur, Nick Greenia (IRS)

10:45 Open Discussion

11:15 An Overview of the Role of Data Sharing in the Production of Federal Statistics
Steven Landefeld, Dennis Fixler (BEA)

11:45 Open Discussion

12:00 pm Lunch and Discussion

Recent Developments in Data-Sharing Regulations and Legislation
Katherine Wallman (OMB)

Research and Policy Perspectives on the Benefits of Data Sharing

1:00 Introduction to the Session
Charles Shultze

1:15 Monetary Policy and Research at the Federal Reserve
Carol Corrado (Federal Reserve Board)

1:45 Open Discussion

2:00 Productivity and Real Output Measurement
Dale Jorgenson (Harvard University)

2:30 Open Discussion

2:45 Break

3:00 Tax Data Needs for Improving Estimates of Corporate Profits
George Plesko (University of Connecticut)

3:30 Open Discussion

3:45 The Importance of Accurate Data and Data Sharing for Budget Forecasting
Douglas Holtz-Eakin (CBO)

4:15 Open Discussion

Conclusions: Data Sharing, Data Quality, and Informed Policy

4:30 Key Points from the Presentations: Directions for the Future
Robert Parker (federal statistics consultant)

Concluding Remarks from the Statistical Agencies
Kathleen Utgoff (BLS); Howard Hogan (Census Bureau); Rosemary Marcuss (BEA)

5:30 Adjourn

ATTENDEES

(* indicates a presenter)

Thomas Anderson, Bureau of Economic Analysis
Stephen Andrews, Bureau of Economic Analysis
Anthony Apostolides, Bureau of Transportation Statistics
Lisa Blumerman, Census Bureau
Ruth Bramblett, Bureau of Economic Analysis
Stephanie Brown, Census Bureau
Carolyn Carroll, STAT TECH, Inc.
Chris Chapman, Bureau of Labor Statistics
Constance Citro, National Research Council
Carol Corrado, Federal Reserve Board*
Ruth Detlefsen, Census Bureau
Chris Ellis, RTI International
Joel Elvery, Bureau of Labor Statistics
Vincent Fang, MacroSys Research and Technology
Daniel Feenberg, NBER (steering committee member)
Jason J. Fichtner, Joint Economic Committee
Dennis Fixler, Bureau of Economic Analysis*
Lucia Foster, Census Bureau
Charles Funk, Census Bureau
John Galvin, Bureau of Labor Statistics
Gerald Gates, Census Bureau
John Golmant , Administrative Office of the U.S. Courts
Nancy M. Gordon, Census Bureau
Nick Greenia, Internal Revenue Service*
Keith Hall, Department of Commerce
John Haltiwanger, University of Maryland (steering committee member)
Charlie Han, Bureau of Transportation Statistics
Paul Hanczaryk, Census Bureau*

Brian A. Harris-Kojetin, Office of Management and Budget
Maurine Haver, Haver Analytics
Brian Higginbotham, Joint Economic Committee
Tina Highfill, Bureau of Economic Analysis
Paul L. Hsen, Census Bureau
Howard Hogan, Census Bureau*
Brian Holly, Census Bureau
Douglas Holtz-Eakin, Congressional Budget Office*
Ned Howenstine, Bureau of Economic Analysis
Ron Jarmin, Census Bureau
Dale Jorgenson, Harvard University*
Arthur Kennickell, Board of Governors of the Federal Reserve System
Sanders Korenman, Baruch College, CUNY
Meredith Krug, Federal Reserve Board
Caryn Kuebler, National Research Council
Herbert Lacayo Jr., Environmental Protection Agency
Steven Landefeld, Bureau of Economic Analysis*
Timothy (Bogong) Li, Bureau of Labor Statistics
Juanita Tamayo Lott, Census Bureau
Christopher Mackie, National Research Council
Rosemary Marcuss, Bureau of Economic Analysis*
Sandra Mathieson, Census Bureau
Mark Mazur, Internal Revenue Service*
Thomas Mesenbourg, Census Bureau
Harry Meyers, National Academy of Public Administration
Kimberly Moore, Census Bureau
Brent Moulton, Bureau of Economic Analysis
Patrick Mullen, Government Accountability Office
Sue Okubo, Bureau of Economic Analysis
Robert Parker, federal statistics consultant*
Thomas Petska, Statistics of Income, Internal Revenue Service
George Plesko, University of Connecticut*
Thomas Plewes, National Research Council
Dylan Rassier, Bureau of Economic Analysis
Ralph Rector, The Heritage Foundation (steering committee member)
Brooks Robinson, Bureau of Economic Analysis
Ruth Runyan, Census Bureau
Ken Ryder, National Academy of Public Administration
Hossain Sanjani, Bureau of Transportation Statistics
Charles Schultze, Brookings Institution (steering committee member)
Molly Shannon, Bureau of Labor Statistics
Gerald Silverstein, Treasury Department
Michael Siri, National Research Council

Timothy Slaper, Joint Economic Committee
Jim Spletzer, Bureau of Labor Statistics*
Bryce Stephens, National Academy of Public Administration
Miron Straf, National Research Council
Ovuokerie Ubogu, Temple University
Kathleen Utgoff, Bureau of Labor Statistics*
Katherine Wallman, Office of Management and Budget*
Diane Willimack, Census Bureau
Ewen Wilson, Census Bureau
Henry Wulf, Census Bureau
Qingshu Xie, MacroSys Research and Technology

Appendix B

Recent Legislation Governing Data Sharing and Access to Federal Tax Data

The Confidential Information Protection and Statistical Efficiency Act (CIPSEA) was passed as Title V of the E-Government Act of 2002 (P.L. 107-347). Enactment of CIPSEA culminated more than 30 years of efforts to standardize and strengthen legal protections for data collected for statistical purposes by federal agencies while permitting limited sharing of individually identifiable business information among three statistical agencies for efficiency and quality improvement.[1]

Title V has two subtitles. Subtitle A, Confidential Information Protection, strengthens and extends confidentiality protection for all statistical data collections of the U.S. government. For all data furnished by individuals or organizations to an agency under a pledge of confidentiality for exclusively statistical purposes, it provides that the data will be used for statistical purposes only and will not be disclosed in identifiable form to anyone not authorized by the title. It makes knowing and willful disclosure of confidential statistical data a class E felony with fines up to $250,000 and imprisonment for up to five years.

Subtitle A pertains not only to surveys, but also to collections by a federal agency for statistical purposes from administrative records (e.g., state government agency records). Data covered under subtitle A are not subject to release under a Freedom of Information Act request. Guidance from the Office of Management and Budget, which is charged to oversee

[1]This overview of CIPSEA is reproduced from National Research Council (2005).

and coordinate the implementation of CIPSEA, is under development. It is intended to cover such topics as the steps that agencies must take to protect confidential information; wording of confidentiality pledges in materials that are provided to respondents; steps that agencies must take to distinguish any data or information they collect for nonstatistical purposes and to provide proper notice to the public of such data; and ways in which agents (e.g., contractors, researchers) may be designated to use individually identifiable information for analysis and other statistical purposes and be held legally responsible for protecting the confidentiality of that information.

Subtitle B of CIPSEA, Statistical Efficiency, permits the Bureau of Economic Analysis (BEA), the Bureau of Labor Statistics (BLS), and the Census Bureau to share individually identifiable business data for statistical purposes. The intent of the subtitle is to reduce respondent burden on businesses; improve the comparability and accuracy of federal economic statistics by permitting these three agencies to reconcile differences among sampling frames, business classifications, and business reporting; and increase understanding of the U.S. economy and improve the accuracy of key national indicators, such as the national income and product accounts.

A key limitation of CIPSEA is that data in the Census Bureau's Business Register, which is constructed in large part out of Internal Revenue Service (IRS) tax data, cannot be shared. New legislation would be required to amend Title 26, which governs access to IRS tax data by BLS, BEA, and other agencies. All access to federal tax information, even within IRS, must be authorized by statute, meaning that legislation has been codified as part of Title 26 of the Internal Revenue Code. Title 26, Section 6103, is the primary law articulating IRS's data-sharing constraints. Section 6103 provides that tax returns and return information are confidential and may not be disclosed by the IRS to other government agencies or employees except as provided in the code. It further specifies non-IRS access, indicating which agencies (or other entities) may have access to tax return information. Congress must enact legislation to add users. Accompanying regulations clarify the purposes for which access may be granted, detailing specific items that can be shared and under what conditions the information will be received. The assistant secretary for tax policy of the Treasury Department is responsible for setting regulations. As with all users, BEA access to IRS tax data is codified in Section 6103. For example, Section 6103(j) specifies (in the accompanying regulations) BEA access to Statistics of Income sample files of corporations; Section 6103(n) grants access to some extracts for partnerships and sole proprietorships. In some cases, policy agreements (such as the Census Bureau-IRS Criteria Agreement) may supplement statute and regulations. Because the confidentiality of tax data is considered crucial to voluntary compliance, if agencies want to establish new efforts to use taxpayer information, executive branch policy calls for a business case to support sharing tax data.

The papers by Nick Greenia and Mark Mazur and by Robert Parker in this volume provide further details of statistical agency access to IRS data. The full text of the CIPSEA legislation can be found at <http://www.eia.doe.gov/oss/CIPSEA.pdf>.

REFERENCES

National Research Council

2005 *Principles and Practices for a Federal Statistical Agency, Third Edition.* Committee on National Statistics, M.E. Martin, M.L. Straf, and C.F. Citro, eds. Division of Behavioral and Social Sciences and Education. Washington, DC: The National Academies Press.

U.S. Congress

2002 *Confidential Information Protection and Statistical Efficiency Act of 2002.* (HR4528). Available: http://www.eia.doe.gov/oss/CIPSEA.pdf [accessed May 2006].

Part II

Background Papers

5

Data-Sharing History and Legislation: Background Notes

Robert P. Parker
Consultant on Federal Statistics

During the past 20 years, the National Research Council of the National Academies has prepared several reports on data sharing and related issues, such as privacy and confidentiality. However, as noted in the discussion here of each of these reports, the topics covered by the Workshop on the Benefits of Interagency Business Data Sharing differ from those in the previous reports. The workshop focused on data and information on businesses, not on individuals, and their use for statistical, not research, purposes. In addition, the workshop focused on sharing by the Bureau of Labor Statistics (BLS), the Bureau of Economic Analysis (BEA), and the Census Bureau; these three statistical agencies are authorized to share data under provision of Subpart B of the Confidential Information Protection and Statistical Efficiency Act (CIPSEA) as enacted in December 2002.

This chapter reviews the history of CIPSEA and of efforts by the Office of Management and Budget (OMB) for related legislation that would expand data sharing by changing the provisions of the Tax Reform Act of 1976, which allow the Internal Revenue Service (IRS) to share tax information with selected statistical agencies. CIPSEA provides statutory confidentiality protection for data collected by statistical agencies for statistical purposes (Subpart A), and it permits identifiable business records to be shared for statistical purposes by BLS, BEA, and the Census Bureau (Subpart B). CIPSEA, however, did not change the provisions of Internal Revenue Code Section 6103, which precludes these agencies from sharing tax return information for statistical purposes. Such additional sharing would allow the agencies to develop a single business register for all of their economic surveys and programs and to use information from the register to improve the quality of their surveys. It would also create additional opportunities to create new statistical programs from existing data.

There have been efforts by several administrations to change Section 6103 to allow the sharing of basic identifying information from IRS records for statistical purposes. This chapter also reviews the positions on this change by the U.S. Government Accountability Office (GAO), the Department of the Treasury, the IRS, and the Joint Committee on Taxation (JCT), as well as the positions of businesses.

Some of the major ongoing data-sharing efforts by the three agencies are reviewed in this chapter, which covers efforts allowed under Section 6103, the International Investment and Trade in Services Act of 1990, and BEA, BLS, and the Census Bureau confidentiality restrictions.

In this chapter, "data sharing" is defined as the exchange of information collected from businesses and individuals or reported to the IRS in identifiable form for statistical purposes. For other key terms, the chapter follows the definitions in CIPSEA. Business data are "operating information and financial data and related information about businesses, tax exempt organizations, and government entities." Identifiable form means information "that permits the identity of the respondent to whom the information applies to be reasonably inferred by either direct or indirect means." Statistical purposes are "the description, estimation, or analysis of the characteristics of groups, without identifying the individuals or organizations that comprise such groups." They also include methods and procedures related to the "collection, compilation, processing, or analysis" of data about these groups and the development of related "measurement methods, models, statistical classifications, or sampling frames." Using definitions from the report of the Privacy Protection Study Commission (1977), the term "research" is used to refer to "any systematic, objective process designed to obtain new knowledge, regardless of whether it is 'pure' (aimed at deriving general principles) or 'applied' (aimed at solving a specific problem or at determining policy)." The term "statistics" in this chapter refers "both to the data obtained through enumeration and measurement and to the use of mathematical methods for dealing with data so obtained."

PREVIOUS REPORTS ON DATA SHARING

Most of the previous reports of the National Research Council on data sharing and related topics touched on issues important to the discussion of changing Section 6103, such as the trade-off between greater access and reduced confidentiality. In general, the findings and recommendations of these reports did not focus on sharing of data and information on businesses or on data sharing for statistical purposes. Although the most recent of these reports was completed after the enactment of CIPSEA, there was little discussion of the data-sharing part of the act, largely because CIPSEA was limited to the data sharing of business data for statistical purposes, not individual data and not for research purposes. In this section, each of these reports is reviewed in terms of its relevance to business data sharing.

In *Sharing Research Data* (National Research Council, 1985), the focus is mostly on what researchers should do. It thus has little relevance to sharing of business data for statistical purposes. The report does include a section that identifies several benefits related to sharing business data. These benefits are the verification of results, improvements in measurement and data collection methods, and the development of theoretical knowledge. In addition, one of the papers prepared for the conference that led to the report notes the creation of new data sets as a benefit of data sharing. However, there is no mention of the benefits of data sharing to statistical agencies, such as the ability to develop common business registers.

Sharing Research Data cites three earlier reports related to data sharing: (1) *Sharing of Social Science Research Data: An Exploratory Conference Convened by the Committee on National Statistics* (National Research Council, 1980); (2) *Setting Statistical Priorities* (National Research Council, 1976); and (3) *Personal Privacy in an Information Society* (Privacy Protection Study Commission, 1977). The first report deals primarily with health data, and the second does not deal at all with data sharing. The Privacy Protection Study Commission report is discussed later in this chapter.

Private Lives and Public Policies: Confidentiality and Accessibility of Government Statistics was published in 1993 (National Research Council, 1993). The Panel on Confidentiality and Data Access, jointly formed by the Committee on National Statistics and the Social Science Research Council, developed recommendations to assist statistical agencies in their stewardship of data for policy decisions and research. The panel met between November 1989 and January 1992 and used the results of related workshops: the Longitudinal Retirement Workshop (September 1987), Confidentiality of and Access to Doctorate Records (November 1988), and Confidentiality and Access to National Center for Education Statistics (January 1991).

The panel sought to take into account the trade-offs between data protection and data access and made recommendations to "enhance data access without decreasing data protection and increase data protection without reducing data access" (p. 19). The panel decided to focus primarily on major federal statistical programs and to cover "organizational," or business, data as well as individual data. This expansion reflected input from statistical agencies, the experience of panel members, and current events. As a result, the report covers major issues of data protection and access to business data; it includes a chapter on a framework for the treatment of these data (Chapter 7) and covers businesses in other chapters that deal with legislation (especially Chapter 5).

With regard to business data, the panel presents several relevant findings and recommendations. The Executive Summary includes a section on "Barriers to Data Sharing Within Government," in which the panel reported (p. 6):

> Some of the laws that govern the confidentiality of statistical data prohibit or severely limit interagency sharing of data for statistical purposes. Laws that control access to administrative records, such as reports of earnings covered by Social Security, restrict their use for statistical purposes. These barriers to data sharing for statistical purposes have led to costly duplication of effort and excessive burden on individuals and organizations who are asked to supply information. They have also made it difficult or impossible to develop data sets needed for policy analysis on topics of major interest to the public.

The panel recommends that barriers to sharing data for both persons and businesses for statistical purposes should be removed subject to strict controls to protect confidentiality. Recommendation 4.1 includes a statement that "additional data sharing should only be undertaken in those instances in which the procedures for collecting the data comply with the panel's recommendations for informed consent or notification" (p. 6). Recommendation 5.1(f) states that "a provision that permits data sharing for statistical purposes under controlled conditions be included in the consistent set of statutes and regulations governing the maintenance of federal statistical records" (p. 6). Recommendation 7.4 reflects the panel's proposal that interagency sharing of data for statistical purposes should include the sharing of lists of businesses by federal and state agencies. Recommendation 7.5 asks for new legislation to expand confidentiality to records collected by all statistical agencies (p. 226); some of this recommendation was subsequently accomplished by CIPSEA.

Although this report cites several examples of reduced sharing as a result of the provisions of Section 6103, there is no explicit recommendation to amend this act. Nevertheless, the panel clearly expresses its views in the following section title in Chapter 7: "Inability to Share Business Lists: An Embarrassment to the Federal Statistical System."

In addition, the report includes several important discussions relating to sharing business data that appear to conflict with the current IRS interpretation of Section 6103. In a discussion on the Census Bureau use of tax data, it states (p. 192):

> For a particular establishment or employer whose identity was originally supplied by the IRS, the Internal Revenue Code allows the Census Bureau to contact the taxpayer, and any response returned to the Census Bureau is considered to be data collected under the authority of the Census Bureau (Title 13 U.S.C.) rather than tax return information.

The report also notes that there is little or no sharing of business data collected by the Census Bureau (p. 192):

> Exceptions are the occasional correction or updating of SIC [Standard Industrial Classification] codes on lists provided by other agencies, under the authority of a 1953 opinion issued by Attorney General McGranery (41 Op. A.G. 120), and the release of certain SSEL [Standard Statistical Establish List] information to BEA, which has become possible as a result of legislation passed by the 101st Congress (Foreign Direct Investment and International Financial Data Improvements Act of 1990, P.L. 101-533). As part of the same legislation, BEA is required to share with BLS and the Census Bureau selected data on foreign direct investment that it collects from business enterprises.

Improving Access to and Confidentiality of Research Data: Report of a Workshop was published in 2000 (National Research Council, 2000). This workshop report does not discuss sharing business data. The report states that a central objective of the workshop was to review the benefits and risks associated with public-use research data files and to explore alternative procedures for restricting access to sensitive data, especially longitudinal survey data that have been linked to administrative records. It focuses on microdata records for research purposes. The workshop did not produce recommendations or specific findings.

Principles and Practices for a Federal Statistical Agency, Third Edition, was published in 2005 (National Research Council, 2005a). Although the report deals with a wide range of activities by statistical agencies, it lists the sharing of microdata records as a "way to improve data quality and develop new kinds of information . . . for statistical agencies that collect similar information" (p. 26). It notes that the sharing of business data would "make it possible to evaluate reporting errors and the completeness of coverage of business firms in different surveys," and that "such sharing would also make it possible to develop more useful and accurate statistics on the nation's economy while decreasing the reporting burden on business data providers" (p. 26).

The report discusses the enactment of CIPSEA, noting that the first formal proposal for data sharing under CIPSEA, which involved matching data from BEA's international investment surveys with data from the Census Bureau's Survey of Industrial Research and Development conducted for the National Science Foundation, was announced in June 2003. (Additional details on this match are discussed later in this chapter.)

Expanding Access to Research Data: Reconciling Risks and Opportunities was published in 2005 by the Panel on Data Access for Research Purposes, which was convened at the request of the National Institute on Aging (National Research Council, 2005b). The panel report, which is built largely on the earlier efforts discussed above, primarily focuses on data for research purposes. The Introduction states: "Thus, there are questions about how to provide researchers—inside and outside government—access to data that can both inform public policy and protect the privacy of respondents and the confidential nature of the information they provide" (p. 8). Consequently, the report is almost exclusively research oriented although it does cover some research using business records. It focuses on microdata research and notes the policy and cost benefits for data linking and sharing. Although not an explicit recommendation, the report does note the need to revise Section 6103, stating (p. 35):

> [It] offers recommendations that, if implemented, will continue the past record of simultaneous improvement along both dimensions. Such improvement will require strong partnership between the research community and statistical and research agencies in the design of innovative research on disclosure avoidance techniques and data access modalities and in the implementation of the advances that result from such research.

The final chapter of the report makes 19 recommendations covering such areas as documentation, planning for and expanding access to public-use files and research data centers, informed consent, and safeguarding confidentiality. None of the recommendations in these areas is directly relevant to the sharing of business records.

In a section on the legal environment, a discussion of CIPSEA points to the need for changing Section 6103. The report notes (p. 23):

> A key element in the Census Bureau's data is its business register, which is constructed with data from the Internal Revenue Service (IRS). However, without new legislation (to amend Title 26 of the U.S. Code, which governs access to IRS tax data), the business register and associated data cannot be shared with BEA and BLS.

The report discusses policy research but limits the discussion to the use of microdata. It states: "Detailed microdata permit in-depth analyses of socioeconomic trends and their antecedents and consequences" (p. 40) and continues with "Detailed microdata are also needed for modeling economic decisions and other kinds of social behavior" (p. 41). Examples of such research noted in the report include research to model savings behavior and firm behavior on pollution abatement. It also includes a section on linking survey and administrative record data, but examples are limited and omit any discussion of BEA's foreign direct investment data. It reports on the 1973 Current Population Survey–Social Security

Administration–IRS (CPS–SSA–IRS) "exact-match file" that was the basis for a major dynamic microsimulation model of social welfare policies and retirement income and for an analysis of the quality of income reporting in the March CPS. The report does not cite the continuing use by the Census Bureau of linkages of survey data and administrative records to assess and improve data quality. It also does not report that BEA and IRS have used the aggregate results of this match program to estimate the extent of nonfiling by unincorporated enterprises. In fact, BEA continues to fund an updated version of this match, but without the SSA records.

Among the papers prepared for this panel, "Privacy, Confidentiality, and Data Sharing" by David McMillen is the most relevant to sharing of business data. It provides an excellent history of CIPSEA, tracing its most recent roots to the Clinton administration's efforts to enact confidentiality and data-sharing legislation in 1995. (See additional details in the next section.) McMillen also identifies several examples of sharing both business and individual records. For matching of information collected under a mandatory authority, he strongly sees the need to notify respondents when the data they report are to be linked to IRS records or made available to another agency. The report indicates some discomfort with this part of McMillen's views and does not incorporate them into the panel's recommendations.

TAX REFORM ACT OF 1976

To provide an understanding of the changes incorporated in 1976 into Section 6103, it is important to first review the developments that led to these changes. (Most of the following discussion is based on a 1983 paper by Wilson and Smith and various editions of the Census Bureau series on the history of the economic censuses.) The most important change was that, prior to the Tax Reform Act of 1976, the president had the authority to issue executive orders permitting access to tax records. The revised Section 6103 limited access to tax records to specific federal and state and local government agencies and only for specific types of tax information. Under the previously issued executive orders, access for statistical purposes had been granted to the Commerce Department, covering both the Census Bureau and the predecessor agencies to BEA (1944 E.O. 9499 and 1961 E.O. 10911), the Securities and Exchange Commission and Federal Trade Commission (FTC) (1959 E.O. 10814 and 1961 E.O. 10908), and the Department of Agriculture (1973 E.O. 11697 and 11709). Under the Tax Reform Act of 1976, access by the Commerce Department was continued, but BEA's access was limited to selected records of corporations, and access by the FTC to conduct the Quarterly Financial Report was continued.

Access by the Department of Agriculture was not continued, although it was restored in the 1990s, when the Census of Agriculture was transferred from the Census Bureau to the Department of Agriculture. In addition to changes in access, the 1976 version of Section 6103 included stricter disclosure rules.

According to *Personal Privacy in an Information Society*, a report by the Privacy Protection Study Commission, a commission mandated by the Privacy Act of 1974, the changes resulted from an increase in disclosure by the IRS during the first half of the 20th century (Privacy Protection Study Commission, 1977). Most of these new uses were authorized without any significant public discussion and, when made public, were justified by the federal and state agency recipients of the data as essential to an essential government function. The commission stated (p. 27 of Appendix 2):

> The abuses that inevitably resulted were from time to time brought to the attention of the Congress and the public, sometimes dramatically. The Nixon Administration allegedly used tax returns to harass its political adversaries, and an announcement early in the 1970's that information about individual taxpayers would be made available to the Department of Agriculture to aid in statistical analysis aroused intense controversy. Allegations that special powers of the Internal Revenue Service were being misused to collect information for purposes well beyond tax administration eventually led to a series of Congressional hearings on the propriety of various uses of tax administration.

This report had been limited in scope to individually identifiable information about individuals, so the commission did not study issues regarding disclosure of information about business entities. Nevertheless, the commission generally agreed with the limited access to tax return information included in Section 6103 in the Tax Reform Act of 1976. (The final report was published in 1977, but the commission had provided its recommendations to Congress in June 1976.)

EFFORTS TO EXPAND DATA SHARING

Before 1976

In a 1979 study, GAO reported on the status of efforts of the Census Bureau to create a centralized business listing, or business register, that would be used by the Census Bureau and other statistical agencies to select samples for various surveys (U.S. General Accounting Office, 1979). By selecting samples from the same register, the list would improve the comparability of the data collected in these surveys regardless of the collecting agency and would reduce duplication in the construction of these lists. The development of such a list had been recommended by various committees and commissions dating back to 1937. In 1968, OMB designated the Census Bureau as the focal agency to develop and maintain the list data. From 1972 to 1978, Congress provided the Census Bureau almost $16 million to develop the list. In 1975, the list became operational and was used by the Census Bureau for the 1977 economic census and current economic survey programs. However, the list could not be shared with other agencies, because the Census Bureau had yet to propose specific legislation to change Title 13 of the U.S. Code to allow it to share the list with other statistical agencies.

The enactment of the Tax Reform Act of 1976 and replacement of previous executive orders with Section 6103 had a major impact on this effort for the centralized business register. Prior to the act, it had been thought that only Title 13 needed to be changed to allow for sharing of the list. However, the act made it clear that because of the extensive use of tax return information by the Census Bureau (as discussed below), it also would be necessary to change Section 6103. In December 1978, the secretary of the treasury informed the secretary of commerce that the Treasury Department would not support legislation to change Section 6103. The secretary stated the reasons for this opposition as follows (p. 56):

> We would, however, have significant concerns with any proposal to amend Section 6103 of the Internal Revenue Code to authorize disclosure of tax return information to be used for SSEL purposes. . . . The considerations which underlie the strict limitations imposed by Congress upon the disclosure of tax return information are the taxpayer's expectation of and right to privacy, and the potential impact of disclosure upon the continuation of compliance with our voluntary tax assessment system.

1983 to 1994

After the unsuccessful effort by the Census Bureau to change Title 13 and Section 6103 to allow it to share its business register with other statistical agencies, OMB sought legislation to allow more extensive data sharing, including the sharing of data on individuals. The new effort would include allowing for sharing by more statistical agencies, but it also would provide statutory protection for data collected by all statistical agencies similar to the protection afforded to the Census Bureau by Title 13. Some agencies had no such supporting legislation, while others, under certain circumstances, were to turn over data collected for statistical purposes to other agencies for nonstatistical purposes. Thus, the proposed legislation would protect data collected by all statistical agencies for statistical purposes and allow these agencies to share these data with other statistical agencies.

The renewed effort to change Section 6103 began in 1983 when the administration sought to have the Congress enact so-called statistical enclave legislation, which would provide a single set of confidentiality policies for all federal agencies and their components that collect data for statistical purposes. This legislation, unlike CIPSEA, would have allowed access to tax records by all statistical agencies and included records of both individuals and businesses. As described by Wilson and Smith (1983), this legislation was to provide "(a) a statutory basis for the traditional promise of confidentiality long given respondents to statistical collections, and (b) restricted sharing of individually identifiable records ('protected statistical files') for exclusively statistical agencies ('protected statistical centers') whose confidential records are provided statutory protection under this legislation" (p. 601). The authors express their personal agreement with the portion of the bill that would ensure that data collected for statistical purposes will be used only for statistical purposes. They also report on the objections of the Treasury Department to increasing access to individually identifiable data for statistical purposes. The authors report that "this aspect of the draft Bill is unacceptable in the opinion of the Department of Treasury because, among other things, it might greatly expand the dissemination of tax return information, attenuate the ability to control such dissemination, and provide insufficient safeguards to prevent further disclosure and erosion of confidentiality." Interestingly, the authors also report that "there also appears to be no provision in the Bill that would aid the research work needed to manage tax policy or tax administration situations. In particular, no data-sharing arrangements for IRS statistical purposes are proposed. All the 'sharing' is in the other direction." It is not clear whether this is the author's objection, the department's, or both.

The authors also make it appear that the basis for the Treasury Department's opposition was that "even though controls may be provided as to redisclosure of information, if the perception is created that tax return information is widely available to others, that perception alone could be very destructive of public confidence in the security and privacy of information provided under the tax administration system, and ultimately destructive of voluntary compliance" (p. 600).

1995 to CIPSEA

The next, and most recent, effort by OMB to increase data sharing and change Section 6103 began in 1995, when the Clinton administration sent to Congress a new set of legislation proposals. This legislation was introduced on a bipartisan basis in the House of Representatives in early 1996.

Under the legislation, eight federal agencies were designated as statistical data centers: BEA, the Census Bureau, BLS, the National Agricultural Statistics Service, the National Center for Education Statistics, the National Center for Health Statistics, the Energy Consumption Division of the Energy Information Administration, and the Science Resources Statistics Division of the National Science Foundation. Enactment of this legislation would have improved the efficiency of the federal statistical system, reduced reporting burden on the public, and improved the quality and usefulness of the federal statistics for economic and social policy decisions.

A key feature of the proposed legislation required that data or information acquired by an agency for purely statistical purposes could be used only for statistical purposes and could not be shared in identifiable form for any other purpose without the informed consent of the respondent. This proposal, now the Statistical Efficiency Act of 1999 (H.R. 2885), was passed by the House in October 1999. It increased statutory protections for the confidentiality of statistical data and permitted sharing of data for statistical purposes among designated agencies. The proposal also called for companion legislation to make complementary changes to provisions set forth in the Statistical Use section of the Internal Revenue Code. According to OMB, the complementary proposal was endorsed by the Treasury Department.

To meet congressional concerns about the types of data to be exchanged under this proposed complementary legislation, the administration proposed a revised version of the bill. Under this version, data sharing would be restricted to BEA, BLS, and the Census Bureau, and only data on businesses would be shared. Finally, at the end of 2002, Congress passed the Confidential Information Protection and Statistical Efficiency Act (CIPSEA) of 2002 as Title V of the E-Government Act (Public Law 107–347). This legislation established a uniform set of safeguards to protect the confidentiality of individually identifiable information acquired from the public for statistical purposes, as well as consistently strong criminal penalties for inappropriate disclosure of such information. The legislation reaffirmed that pledges of confidentiality would be honored and gave additional weight and stature to policies that statistical agencies have pursued for decades, assuring respondents who provide statistical information that their responses will be held in confidence and will not be used against them in any government action. CIPSEA also authorizes the sharing of business data among the BEA, BLS, and the Census Bureau. The companion legislative proposal, which would make complementary changes on access to tax return data by these agencies, has been endorsed by the Treasury Department and submitted to Congress; the proposal has not been enacted.

CIPSEA

In 2002, Congress passed CIPSEA. The following is the summary of the legislation from *Principles and Practices for a Federal Statistical Agency, Third Edition* (National Research Council, 2005a, pp. 62-63):

> Enactment of CIPSEA culminated more than 30 years of efforts to standardize and strengthen legal protections for data collected for statistical purposes by federal agencies while permitting limited sharing of individually identifiable business information among three statistical agencies for efficiency and quality improvement.
>
> Title V has two subtitles. Subtitle A, Confidential Information Protection, strengthens and extends confidentiality protection for all statistical data collections of the U.S. government. For all data furnished by individuals or organizations to an agency under a pledge of confidentiality for exclusively statistical purposes, it provides that the data will be used only for statistical purposes and will not be disclosed in identifiable form to anyone not authorized by the title. It makes knowing and willful disclosure of confidential statistical data a class E felony with fines up to $250,000 and imprisonment for up to 5 years. Subtitle A pertains not only to surveys, but also to collections by a federal agency for statistical purposes from administrative records (e.g., state government agency records). Data covered under Subtitle A are not subject to release under a Freedom of Information Act request. Guidance from OMB, which is charged to oversee and coordinate the implementation of CIPSEA, is under development. It is intended to cover such topics as the steps that agencies must take to protect confidential information; wording of confidentiality pledges in materials that are provided to respondents; steps that agencies must take to distinguish any data or information they collect for nonstatistical purposes and to provide proper notice to the public of such data; and ways in which agents (e.g., contractors, researchers) may be designated to use individually identifiable information for analysis and other statistical purposes and be held legally responsible for protecting the confidentiality of that information.
>
> Subtitle B of CIPSEA, Statistical Efficiency, permits the Bureau of Economic Analysis (BEA), the Bureau of Labor Statistics (BLS), and the Census Bureau to share individually identifiable business data for statistical purposes. The intent of the subtitle is to reduce respondent burden on businesses; improve the comparability and accuracy of federal economic statistics by permitting these three agencies to reconcile differences among sampling frames, business classifications, and business reporting; and increase understanding of the U.S. economy and improve the accuracy of key national indicators, such as the national income and product accounts.

Post-CIPSEA

Since the enactment of CIPSEA, OMB has been working again to get the companion legislation to change Section 6103 enacted by Congress. In *Statistical Programs of the U.S. Government: Fiscal Year 2005*, OMB reported on the latest developments in the effort to change Section 6103. In a section on "Statistical Confidentiality and Data Sharing," OMB reported the following (p. 44):

> A companion legislative proposal would make complementary changes to provisions set forth in the "Statistical Use" section of the Internal Revenue Code. These changes would represent the first major revision of these policies in more than 20 years, reducing the amount of sensitive tax information that will change hands to support statistical programs while substantially increasing the effectiveness of that support. This objective would be achieved by carefully defining statistical needs and taking advantage of the efficiencies that can be achieved by modern sampling methods. The complementary proposal has been endorsed by the Treasury Department and submitted to the Congress.

Positions on Changing Section 6103

This section describes the public positions of various organizations with an interest in data sharing, particularly with tax return records. (This section excludes OMB, which as discussed above, has consistently supported changing the tax laws to permit more data sharing.) The views of GAO, the Council of Economic Advisers, the Department of the Treasury and the Joint Committee on Taxation are presented. It also includes the views of a special study commissioned by the Administrative Conference of the United States.

Administrative Conference

In response to a congressional request, the administrative conference in 1975 commissioned a team of tax experts headed by Charles Davenport to study the operations of the IRS. The result was the *Report on Administrative Procedures of the Internal Revenue Service, October 1975, to the Administrative Conference of the United States*, (Davenport, 1976) which included a section on tax return confidentiality. This section reviewed the history and rationale for tax return access for statistical purposes. The report concluded that "it appears that the use of tax data by Census, though not consistent with revenue administration, is a use which can be considered beneficial and is one which does not appear have any undesirable side effects" (p. 880). The study came to a similar a conclusion for BEA but not for the Statistical Research Service (SRS) of the Department of Agriculture. The key distinguishing factor for the commission was that Census and BEA were strictly statistical agencies and did not engage in activities related to other functions of the department. In contrast, the commission determined that the statistics collected by SRS were "for policy making by the agency of which it is a part" (p. 886).

Government Accountability Office

In 1979, GAO issued a report on recent developments in the Census Bureau's efforts to create a centralized business register for use by other statistical agencies (U.S. General Accounting Office, 1979). GAO, which supports changing Section 6103, summarized its findings as follows (title page):

> The Bureau of the Census has developed the Standard Statistical Establishment List, a comprehensive list of businesses in the United States. Many Federal agencies could use such information. But confidentiality laws prevent the Census Bureau from sharing List information with other agencies.
>
> Amendments to these laws would help improve the quality and comparability of economic statistics and reduce business response burden from numerous Federal statistical surveys.
>
> Because some of the List data comes from the Internal Revenue Service, the Treasury Department has reservations about using tax information for statistical purposes. However, the Commerce Department plans to introduce proposals for changes to the confidentiality laws and GAO recommends favorable congressional consideration.

In 1998, GAO testified before Congress on *Statistical Agencies: Proposed Consolidation and Data- Sharing Legislation* (U.S. General Accounting Office, 1998). The testimony was about the data-sharing legislation submitted in 1996 and 1997 that would permit limited sharing of data among designated statistical agencies for statistical purposes, subject to procedural safeguards. The testimony included the following statement of GAO's position (pp. 5-6):

> For the past 2 decades, we and others have urged legislative changes that would allow greater sharing of data and information on data sources among agencies, but so far these efforts have met with little success. The Paperwork Reduction Act of 1980 gave the Director of OMB the authority to direct a statistical agency to share information it had collected with another statistical agency. However, this authority was limited since it did not apply to information that was covered by laws prohibiting disclosure outside the collecting agency. In the early 1980s, the statistical agencies, under OMB's leadership, tried to further enable federal statisti-

> cal agencies to share data. They attempted to synthesize, in a single bill, a set of confidentiality policies that could be applied consistently to all federal agencies or their components that collected data for statistical purposes. This effort, which was known as the "statistical enclave" bill, would have allowed statistical agencies to exchange information under specific controls intended to preserve the confidentiality of the data providers. A bill was introduced in Congress but was not enacted.
>
> More recent proposals concerning data sharing have called for enactment of legislation that would allow statistical agencies to share data and information with appropriate safeguards to protect against breaches of confidentiality. These proposals were not adopted, in part because of general concerns that greater data sharing might endanger the privacy of individuals. Both the Economic Statistics Initiative under President [George H.W.] Bush and the National Performance Review (NPR) under President Clinton have recommended such actions. NPR recommended the elimination of legislative barriers to the exchange of business data among federal statistical agencies, and we agreed with this recommendation. The NPR recommendation did not address the sharing of information on individuals. Some officials of statistical agencies and Members of Congress, however, have argued that a distinction should be made between the sharing of business data and the sharing of personal data about individuals. They noted that breaches of confidentiality protection when personal information is involved may be more serious. The National Academy of Sciences has made recommendations regarding the need for appropriate legislative provisions on data sharing that the Subcommittee may wish to consider in its deliberations on S. 1404.

In 2001, GAO issued *Record Linkage and Privacy: Issues in Creating New Federal Research and Statistical Information* (U.S. General Accounting Office, 2001). This report reviews the background of the key issues related to data sharing. However, the focus on the report was "on linkage projects that involve person-specific data, are conducted under federal auspices (or with federal funding), and produce new research or statistical information" (p. 10). The report did not cover business data sharing.

Council of Economic Advisers

In the 2002 *Economic Report of the President*, the Council of Economic Advisers noted the critical need for reliable data and that data sharing would increase their quality. This excerpt notes the submission of proposals in 1999 and that the Administration would continue to work for their passage (p. 25):

> Recent economic events have emphasized the importance of timely economic information. Thus one area deserving considerable attention is the need for readily accessible real-time data. Investment in sources of these

data could yield handsome dividends, especially at key junctures in the business cycle. . . .

Moreover, the quality of existing statistics is far from perfect and could be enhanced with further investment. Even real GDP, generally thought of as a reliable measure of overall activity in the U.S. economy, is susceptible to considerable revisions. . . . Such revisions lead to uncertainty for both government and private decisionmakers, which can cause costly delays. . . .

A number of steps can be taken to improve the accuracy and timeliness of economic statistics. In particular, targeted improvements to the source data for the national accounts would go a long way toward illuminating the causes of the growing statistical discrepancy. Another cost-effective measure would be to ease the current restrictions on the sharing of confidential statistical data among Federal statistical agencies. Such data sharing, which would be done solely for statistical purposes, is currently hindered by lack of a uniform confidentiality policy. Confidentiality is of key importance to all agencies and to the individuals and businesses who participate in Federal surveys, but a uniform confidentiality policy would allow agencies such as the Bureau of Economic Analysis, the Bureau of Labor Statistics, and the Bureau of the Census to cost-effectively compare and improve the quality of their published statistics while preserving confidentiality. In the past, attempts have been made to pass legislation, together with a conforming bill to modify the Internal Revenue Code, allowing such data sharing under carefully crafted agreements between or among statistical agencies. In 1999 such legislation passed the House but stalled in the Senate. The Administration will continue to seek passage of data-sharing legislation to improve the quality and effectiveness of Federal statistical programs.

Treasury Department and the Joint Committee on Taxation, 2000

Section 3802 of the Internal Revenue Service Restructuring and Reform Act of 1998 requires the secretary of the treasury and the JCT to conduct separate studies of the scope and use of provisions regarding taxpayer confidentiality and to report the findings of such a study, together with any recommendations deemed appropriate to Congress. The JCT published its report on January 28, 2000; the Office of Tax Policy of the Department of the Treasury submitted its report on October 2, 2000.

The JCT and the Treasury Department disagree on changing Section 6103 to permit expanded data sharing. The JCT recommended that "new access to returns and return information should not be provided unless the requesting agency can establish a compelling need for the disclosure that clearly outweighs the privacy interests of the taxpayer" (p. 196). (See below for information on business taxpayer views.) The JCT report did not explain how to determine both the pros and cons of such a decision, nor did it appear to fully accept the need for the existing access. (Both the JCT and the Treasury Department recommended dropping the Federal Trade Commission from access under Section 6103.)

The Treasury Department recommended "the disclosure authority of Section 6103(j) should be expanded to additional specified statistical agencies, and such agencies should be permitted, upon prior Treasury approval, to share IRS data with each other." It also specified the agencies to include and that the change to Section 6103 should cover both individual and business records.

Treasury Department, 2001 to Present

In the 2001-2005 *Statistical Programs of the United States Government*, OMB reports support for the companion legislation that would change Section 6103.

Business Taxpayers

As noted above, the JCT identified the key issues in determining whether to change Section 6103 when it wrote "new access to returns and return information should not be provided unless the requesting agency can establish a compelling need for the disclosure that clearly outweighs the privacy interests of the taxpayer" (p. 196). Although there is a substantial amount of research on individuals' views on confidentiality and data sharing, much of it conducted for the decennial censuses, there is limited information on the views of businesses. The information that is available seems to indicate that business taxpayers are willing to allow more access to statistical agencies for some types of tax return information.

Private business economists have actively supported data sharing. In 1996, Maurine Haver, president of Haver Analytics and chair of the Statistics Committee of the National Association for Business Economics (NABE), testified before the House Subcommittee on Government Management, Information and Technology. In her testimony on consolidating the three major economic statistical agencies (H.R. 2521), she expressed support for the inclusion in that bill of provisions to provide for data sharing among these agencies, as it would allow for the creation of a single business register. In 2001, Richard Berner, the president of NABE and managing director and chief U.S. economist, Morgan Stanley, Inc., testified before the House Subcommittee on the Census Committee on Government Reform on the Statistical Efficiency Act of 1999. In his testimony, Berner reported on NABE's support for the reintroduction of this act, which had been passed unanimously by the House in the previous Congress. He testified "NABE believes that our national data collection efforts should be as efficient as possible. To that end, we believe that Congress should mandate 'data sharing' among the agencies, solely for statistical purposes." He noted that existing confidentiality statutes are barriers to such data sharing because "they virtually guarantee duplication of effort and inconsistencies among related data sets collected by the affected agencies. Moreover, they prevent agencies from undertaking new analyses that could improve the information available to policy makers. This is not a cost-effective way to run any business—either public or private."

A 2001 article called "Business Perceptions of Confidentiality" reported on the results of a survey conducted by the Urban Institute on these business perceptions (Greenia, Lane, and Jensen, 2001). Of particular interest to the issue of privacy interests is the response to a set of questions about types of data that businesses view as very sensitive. The survey results showed that less than 5 percent of respondents thought their name and address and industrial activity were very sensitive, but 85 percent of respondents were very sensitive to data about their employees. Between 50 and 75 percent of the respondents thought their financial data were very sensitive.

CURRENT INTERAGENCY SHARING OF BUSINESS DATA

Among the many examples of interagency sharing of business data, the oldest and most extensive is the use of tax information by the Census Bureau to support its economic statistics programs. With the cooperation of the IRS in modifying questions on tax returns and timely delivery of tax records, the Census Bureau has been able to reduce reporting burden in the quinquennial economic censuses for small businesses, to improve the quality of key data from these censuses, and to develop new statistical programs. This section discusses this sharing as well as other selected examples; it excludes the County Business Patterns report (the Census Bureau, IRS, and SSA) and the sharing of BLS quarterly unemployment insurance data with BEA.

Sharing by the Census Bureau and IRS

As reported in various editions of the history of the economic censuses, the extensive use of tax return records, from both the IRS and the SSA, began with the 1954 economic censuses. In previous censuses, field enumeration was used to collect the data. In 1954, the Census Bureau developed a mailing list from IRS records of employers with the industry classification based on lists from the SSA. In addition, to assist the Census

Bureau, the IRS changed the 1954 tax forms to include information needed to determine the physical location of the business rather than the mailing address and other changes needed by the Census Bureau to determine whether firms had employees. As a result, for the first time, it became possible for the Census Bureau to use tax return information as a source of census data. The Census Bureau decided that because firms with no paid employees represented a very small portion of the volume of retail and service trade but constituted a large number of businesses, it would forgo the collection of certain information from these firms and use the available tax return information to "impute" a complete report. For 1954, they used this technique by selecting a 50 percent sample of retail and service nonemployers. In 1963, this process was improved by the cooperation of the IRS in designing new census coverage questions for the 1963 tax returns, assigning industry classifications to nonemployer returns in the field offices. In 1967, the Census Bureau used all tax records and eliminated the sample. In addition, beginning with the 1987 economic census, a separate publication on nonemployer statistics was introduced. These data were published annually beginning with 1998, and the data for 2003 were released in June 2005.

The use by the Census Bureau of the IRS and SSA records was expanded significantly for the 1967 economic censuses. For 1967, the Census Bureau began using the same information to impute a census report for selected small employers. As a result, about 3 million small establishments were relieved of the task of completing census questionnaires. Overall, data for approximately 60 percent of the establishments included in the scope of the census were obtained through the use of IRS and SSA records. (Although large in number, these small establishments account for only about 7 percent of total sales.) In another expansion of the use of the IRS and SSA records, the Census Bureau also began to use tax return information for the larger employer firms to review the quality of their responses in the censuses.

The Census Bureau also has used its access to tax return information to create a new special supplement to the economic censuses, which is now known as the Survey of Business Ownership (SBO). Using data collected in the economic censuses, business tax return information from the IRS, and administrative data from the SSA, the SBO provides statistics that describe the composition of U.S. businesses by gender, race, and ethnicity, as well as owner's age, education level, and veteran status. In the 1997 economic census, this survey was conducted as two separate surveys; a survey of minority-owned businesses and a survey of women-owned businesses. These surveys have been conducted as part of the economic censuses since 1972.

Sharing by BEA and the IRS

Under access to tax returns provided by executive orders and now Section 6103, BEA uses individual corporate income tax return data to estimate adjustments to the published IRS corporate profits data and converts these IRS data to accounting concepts underlying gross national product. For its regional economic accounts program, using a provision of Section 6103 that allows IRS to give access to BEA to review and tabulate individual tax return records for IRS, BEA produces tabulations (after review by the IRS to ensure taxpayer confidentiality) of nonfarm proprietors' income by state and county that are used to distribute BEA's national totals.

Sharing by the Census Bureau and BLS

Another effort to improve the quality of the economic censuses came in 1992 when the Census Bureau and BLS entered into a memorandum of understanding regarding the use of BLS records by the Census Bureau for the economic censuses. Under this agreement, the Census Bureau compiled a list of single establishment employer identification numbers for which it needed the BLS industrial classification codes. The BLS matched this against its own business register and returned a listing of matched records with the appropriate industrial classification codes. The codes provided by BLS were from its three-year classification update program and provided the Census Bureau with more up-to-date classification information, as the Census Bureau typically updated its classification of these businesses once every five years. Records are not available to determine the extent to which the Census Bureau classifications were changed as a result of this effort.

In 1998, BLS and the Census Bureau entered into another memorandum of understanding. This agreement authorized the agencies to conduct research to evaluate (1) the business registers maintained by the two agencies, (2) opportunities to improve each other's list, and (3) the benefits and risks of sharing list information. After an initial round of comparisons had been completed and a report presented to the Federal Economic Statistics Advisory Committee, the project was stopped by the IRS because the Census Bureau may have given BLS staff access to tax return information embedded in the census register. OMB has reported that this research has resumed under a memorandum of understanding that has been approved by IRS.

Sharing Between BEA and the Census Bureau

The International Investment and Trade in Services Act of 1990 authorized BEA and the Census Bureau to exchange data collected by BEA under this act and data collected by the Census Bureau under Title 13. The purpose of the exchange was to improve the quality of each agency's data and to allow for the preparation of a report that showed the distribution by industry and geography of establishments owned by foreign direct investors. The act authorized a similar exchange between BEA and BLS.

In 2003, using the authority from CIPSEA, BEA and the Census Bureau entered into an arrangement to match data from BEA's surveys called Foreign Direct Investment in the United States and U.S. Direct Investment Abroad with data from the Census Bureau Survey of Industrial Research and Development (SIRD). Working with the National Science Foundation, which funds SIRD, the project demonstrated the feasibility of linking the the Census Bureau and BEA survey data to produce new data on the domestic and international dimensions of U.S. research and development (R&D) activity. The project generated new data on R&D activities of U.S. and foreign multinational companies and on the location of U.S. R&D activity by state. The project also allowed for a higher degree of integration between data on the domestic dimensions of R&D and data on the international dimensions. The project also provided benefits to both the Census Bureau and BEA through improvements in data quality. Research on the outcomes of the matching resulted in the Census Bureau's adding over 500 companies to the sample for the 2003 SIRD and over 60 companies to the 2004 SIRD sample frame. The matching also indicated significant differences in the reporting of both R&D expenditures and industry classification. These differences were resolved and resulted in improved quality in BEA's estimates.

Sharing Between BEA and BLS

BLS produced several reports as a result of the sharing authorized by the International Investment and Trade in Services Act of 1990. In 1993, BLS released 1989 employment data on the occupations for manufacturing by industry and country of ownership in foreign-owned establishments. BLS prepared this report using data from the BEA's 1989 Annual Survey of Foreign Direct Investment in the United States to identify foreign-owned establishments in the BLS Covered Employment and Wages data. These establishments were then linked to establishments from the BLS Occupational Employment Statistics Survey of Manufacturing Industries for 1989.

In 1995, BLS released tabulations for the fourth quarter of 1991 showing a distribution by industry and geography for foreign-owned establishments in the United States based on employment data from their Covered Employment and Wages data. BEA is also planning to provide BLS with company and revenue data from several of its international surveys so that BLS can study expanding their International Price Program to include a new price index for royalties and license fees. BEA hopes to benefit from this sharing by comparing its sample frame with the one compiled by BLS for this price program.

REFERENCES

Berner, R.
2001 Testimony before the House Subcommittee on the Census Committee on Government Reform on the Statistical Efficiency Act of 1999, Washington, DC.

Council of Economic Advisers
2002 *Economic Report of the President*. Washington, DC: Executive Office of the President.

Davenport, C., and the Administrative Conference of the United States
1976 *Report on Administrative Procedures of the Internal Revenue Service, October 1975, to the Administrative Conference of the United States*. Senate document, 94th Congress, 2nd Session; 94-266.

Greenia, N., J. Lane, and B. Jensen
2001 Business perceptions of confidentiality. Pp. 395-429 in *Confidentiality, Disclosure, and Data Access: Theory and Practical Applications for Statistical Agencies*. P. Doyle, J.I. Lane, J.M. Theeuwes, and L.V. Zayatz, eds. Amsterdam, The Netherlands: North-Holland, Elsevier Science.

Haver, M.
1996 Testimony before the House Subcommittee on Government Management, Information and Technology, 104th U.S. Congress, Washington, DC.

Joint Committee on Taxation
2000 *Study of Present-Law Taxpayer Confidentiality and Disclosure Provisions as Required by Section 3802 of the Internal Revenue Service Restructuring and Reform Act of 1998, Volume I: Study of General Disclosure Provisions* (JCS-1-00). Washington, DC: U.S. Government Printing Office.

McMillen, D.
2003 Privacy, Confidentiality, and Data Sharing. Unpublished paper presented at the National Research Council's Committee on National Statistics Confidential Data Access for Research Purposes Workshop, Washington, DC.

National Research Council
1976 *Setting Statistical Priorities: Report of the Panel on Methodology for Statistical Priorities of the Committee on National Statistics*. Assembly of Mathematical and Physical Sciences, Commission on Behavioral and Social Sciences and Education. Washington, DC: National Academy Press.
1980 *Report of an October 1979 Conference on Sharing of Social Science Research Data*. Committee on National Statistics, Commission on Behavioral and Social Sciences and Education. Washington, DC: National Academy Press.

1985 *Sharing Research Data*. Committee on National Statistics. S.E. Fienberg, M.E. Martin, and M.L. Straf, eds. Commission on Behavioral and Social Sciences and Education. Washington, DC: National Academy Press.

1993 *Private Lives and Public Policies: Confidentiality and Accessibility of Government Statistics*. Panel on Confidentiality and Data Access, Committee on National Statistics. G.T. Duncan, T.B. Jabine, and V.A. de Wolf, eds. Commission on Behavioral and Social Sciences and Education. Washington, DC: National Academy Press.

2000 *Improving Access to and Confidentiality of Research Data: Report of a Workshop*. Committee on National Statistics. C. Mackie and N. Bradburn, eds. Commission on Behavioral and Social Sciences and Education. Washington, DC: National Academy Press.

2005a *Principles and Practices for a Federal Statistical Agency, Third Edition*. Committee on National Statistics. M.E. Martin, M.L. Straf, and C.F. Citro, eds. Division of Behavioral and Social Sciences and Education. Washington, DC: The National Academies Press.

2005b *Expanding Access to Research Data: Reconciling Risks and Opportunities*. Committee on National Statistics, Division of Behavioral and Social Sciences and Education. Washington, DC: The National Academies Press.

Office of Federal Statistical Policy and Standards

1978 *A Framework for Planning U.S. Federal Statistics for the 1980's*. Washington, DC: U.S. Department of Commerce.

Privacy Protection Study Commission

1977 *Personal Privacy in an Information Society*. Washington, DC: U.S Government Printing Office.

Singer, E.

2001 The potential of confidentiality and attitudes toward data sharing by federal agencies. Pp. 341-370 in *Confidentiality, Disclosure, and Data Access: Theory and Practical Applications for Statistical Agencies*. P. Doyle, J.I. Lane, J.M. Theeuwes, and L.V. Zayatz, eds. Amsterdam, The Netherlands: North-Holland, Elsevier Science.

Social Security Administration

1991 Social Security Administration's continuous work history sample. *Social Security Bulletin* 52(10).

U.S. Bureau of Economic Analysis and U.S. Bureau of the Census

2003 *Foreign Direct Investment in the United States: Establishment Data for 1997*. Washington, DC: U.S. Government Printing Office.

U.S. Census Bureau

1968-2002 *Economic Censuses*. Washington, DC: U.S. Department of Commerce.

2005a *County Business Patterns: 2003*. Washington, DC: U.S. Department of Commerce.

2005b *Research and Development Data Link Project: Final Report*. Washington, DC: U.S. Department of Commerce.

2005c *Nonemployer Statistics Report: 2003*. Washington, DC: U.S. Department of Commerce.

2005d *2002 Survey of Business Owners: Advanced Report*. Washington, DC: U.S. Department of Commerce.

U.S. Department of Labor, Bureau of Labor Statistics

1993 *New Research on Occupations in Foreign-owned Manufacturing Establishments in the United States*. (News release.) Washington, DC: U.S. Department of Labor.

1995 *Employment and Wages in Foreign-owned Businesses in the United States, Fourth Quarter 1991*. (News release.) Washington, DC: U.S. Department of Labor.

U.S. Department of the Treasury, Department of Tax Policy

2000 *Report to the Congress on Scope and Use of Taxpayer Confidentiality and Disclosure Provisions as Required by Section 3802 of the Internal Revenue Service Restructuring and Reform Act of 1998, Volume I: Study of General Disclosure Provisions.* Washington, DC: U.S. Department of the Treasury.

U.S. General Accounting Office

1979 *After Six Years, Legal Obstacles Continue to Restrict Government Use of the Standard Statistical Establish List.* Washington, DC: U.S. General Accounting Office.

1998 *Statistical Agencies: Proposed Consolidation and Data-Sharing Legislation.* Washington, DC: U.S. General Accounting Office.

2001 *Record Linkage and Privacy: Issues in Creating New Federal Research and Statistical Information.* Washington, DC: U.S. General Accounting Office.

U.S. Office of Management and Budget

2001-2005 *Statistical Programs of the United States Government, Fiscal Years 2002-2005.* Washington, DC: Executive Office of the President.

Wilson, O., and W. Smith

1983 Access to tax records for statistical purposes. Pp. 595-601 in *Proceedings of the Survey Research Methods Section.* Alexandria, VA: American Statistical Association.

6

IRS Data, Data Users, and Data Sharing

Nick Greenia and Mark Mazur
Internal Revenue Service[1]

The importance of tax data to the federal statistical system, in both identifiable and anonymized form, derives from the fact that they are a national asset, a virtual treasure of information. Tax data are rich in both individual and organizational financial details that are useful in a wide variety of situations. First, these data underpin the administration of the federal tax system, which in turn provides the resources for many federal obligations. Second, tax data have almost as important a role as inputs to critical statistical systems that inform analysts and policy makers both inside and outside government. A critical issue for both the tax system and the federal statistical system is the determination of when a compelling need exists for identifiable tax data (often known as federal tax information or FTI) in lieu of aggregate anonymized data. The balance between the sometimes opposing interests of these systems is the focus of this chapter.

TAX DATA COLLECTED

The Internal Revenue Service (IRS) collects data for a variety of entities—covering over 130 million individuals and over 20 million businesses, tax exempt organizations, and governmental entities. The scope of tax return data, often including complete balance sheets and financial

[1]The views expressed in this paper are those of the authors and may not represent the official positions of the Internal Revenue Service or the Treasury Department.

statements, is vast and contains information on everything from net business profits to charitable contributions made by individuals. Moreover, the regularity of the data provided—annually, quarterly, and even monthly for some returns—and the fact that much of the data are captured electronically and for the universe of filers, makes FTI a potent resource for research and analysis.

The subject of business data can be a broad one, covering corporations, partnerships, and sole proprietorships and both employers and nonemployers. Intuition can be a poor guide for the types of data collected and available. For example, employer data are collected for part-time and full-time sole proprietorships, associated with individual tax returns, corporations, and partnerships, as well as entities that are not typically thought of as businesses, such as nonprofit organizations. Employment data themselves can be compiled at the employer level through the employment tax returns filed by businesses (for example the Form 941 series long used by the Census Bureau). They can also be compiled at the employee level and associated with the related employers through Social Security number/employer identification number (SSN/EIN) crosswalks (for example, using the SSNs and EINs captured from Form W-2, used to report annual wage and salary payments).

Typically, the IRS tracks business data at the EIN or enterprise level, but not at the establishment or place of business level unless they are one and the same. This practice differs from that used by most federal statistical agencies. Tax data accuracy is helped by the IRS compliance programs, including legal disincentives for noncompliance. Nevertheless, given the scope and frequency of the data processed by the IRS, the agency cannot ensure the accuracy of all items or the complete (100 percent) coverage of entities. That is, FTI faces limitations similar to those of data sets maintained by statistical agencies, so the tax data system per se should not be viewed as the panacea for statistical program deficiencies. As experience has shown, there will always be gaps and inconsistencies, even in relatively high-quality data sets.

PURPOSE OF DATA COLLECTION

Fundamentally, FTI is collected for use in administering the tax system, including tax policy analysis in the administration (the Department of the Treasury's Office of Tax Analysis) and Congress (the Joint Committee on Taxation). The IRS considers the successful administration of the tax system as highly dependent on voluntary compliance by millions and millions of taxpayers. In turn, voluntary compliance is seen as reliant on the protection—including the perceived protection—of taxpayer data confidentiality. Taxpayers share personal information with the IRS and are assured that their personal data will be handled with the utmost care. The IRS believes that the tax administration purpose of FTI is paramount, and other uses of tax data, including statistical uses, must not interfere with that purpose.

The statistical usage of FTI is authorized by the statute (Section 6103(j) of the Internal Revenue Code, IRC) and associated Treasury regulations, which detail specific items and clarify the purposes for which access by parties outside the IRS may be granted. The tax code implicitly recognizes that statistical and administrative uses share common ground, in that both missions are dependent on high-quality data. In summary, there are two major goals for FTI. First, the data's confidentiality should be protected, so that voluntary compliance and the workings of the tax system are not harmed. Second, the data should be used effectively and efficiently for authorized purposes. It should be clear from these two goals that the role of the IRS with respect to tax data is less one of ownership than stewardship.

DATA USERS

The foremost use of tax data is administering the tax system and includes such functions as taxpayer account processing, audit and other compliance activity, research, and the compilation of statistics. In addition, FTI is provided, through the federal-state program, to state tax agencies in order to assist with states' tax administration needs. In fact, states account for the lion's share of FTI record disclosures to outsiders—in 2004, over 3 billion of the total 4.5 billion reported disclosures. However, the uses of tax data go well beyond that of tax administration, as the nation has long recognized their value not only for the formulation of tax policy and other program uses (such as Social Security) but also for statistical purposes. For the former purpose, tax data are used extensively by Treasury's Office of Tax Analysis and the congressional tax-writing committees—the Joint Committee on Taxation, the Senate Finance Committee, and the House Ways and Means Committee. Other congressional uses include oversight work undertaken for a tax-writing committee, for example, by the U.S. Government Accountability Office.

For a handful of federal entities listed in the tax code, selected identifiable tax data—by no means all items—are provided for statistical purposes. These consist of the Bureau of Economic Analysis (BEA), the Census Bureau, the Department of Agriculture's National Agricultural Statistical Service (USDA-NASS), and the Congressional Budget Office (CBO). The Census Bureau accounts for most of the statistical-purpose FTI record disclosures: over one billion in 2004. Tax data are also broadly used in statistical nonidentifiable form (usually tabulations) to assist other entities, such as businesses, policy think tanks, federal agencies not authorized to receive identifiable data, and academic researchers.

AUTHORIZATION PROCESS FOR ACCESS

Every access to FTI, even within the IRS, must be authorized by statute, meaning that legislation has been codified as part of Title 26 of the United States Code. Furthermore, the statute requires that only the minimum amount of authorized FTI be provided for accomplishing a given authorized task. These constraints apply to all users, including the IRS. For example, virtually all access to FTI within the IRS is authorized for the purpose of tax administration, which is multifaceted, under Section 6103(h)(1) of the IRC.

As might be expected, given the sensitivity of FTI, the law governing access to confidential or identifiable tax data, especially for statistical purposes, is restrictive with respect to both access and use. Thirty years after the tax code was overhauled with the 1976 Tax Reform Act, the number of entities with statistical access to FTI can still be counted on one hand: the Census Bureau, BEA, CBO, and USDA-NASS.[2] Based on the statutory record to date, it seems clear that Congress has regarded any expansion in access to FTI for statistical purposes as deserving of cautious and comprehensive consideration. Unsurprisingly, the rate of change has been glacial, primarily involving USDA-NASS, when the Census of Agriculture was transferred to that agency from the Census Bureau in the late 1990s, and CBO soon after, with its statutory addition for the purpose of long-term modeling of Social Security and Medicare. Even in these instances, however, the historical precedent provided some reassurance regarding the entry of these two new members to the FTI club. Working as special sworn status individuals in the Census Bureau, NASS had conducted much of the previous agriculture censuses. Similarly, CBO had long-standing experience in handling FTI as an agent for the Joint Tax Committee under Section 6103(f)(4) of the IRC. Thus, neither was a novice with regard to either FTI or the associated culture of confidentiality that FTI access requires.

Adding statistical users or increasing access for current users of FTI means Section 6103(j) must be amended; that is, a new law must be passed. Thus, the first requirement of any data-sharing proposal entailing access

---

[2]The Federal Trade Commission's inclusion in this statute is vestigial, as its Quarterly Financial Report function was transferred to the Census Bureau in the mid-1980s. Although Treasury is also listed in the statute, virtually all of its FTI receipts are authorized by Section 6103(h)(1) as being related to tax administration.

to tax data is that the agencies proposed for sharing data must all be in the tax statute or Section 6103(j). This is a necessary but not a sufficient condition, as the agencies must also share statutory authorization to receive the same types of data—for example, corporate total income and individual investment income. In addition, applicable regulations, which require formal approval by the Treasury Department's assistant secretary for tax policy, may also be needed to authorize access and use of the same tax items for the agencies statutorily enabled to receive FTI. Treasury regulations may not only list the specific item content an agency is authorized to receive but also stipulate a more focused purpose. Regulations can also be amended to remove items that are no longer needed by an authorized recipient. In fact, need and, in particular, the requirement of providing only the minimum amount of data needed to accomplish a compelling agency task, is a bedrock principle used for determining not only the necessity of a new statute, but also a regulation amendment. Historically, both Congress and the Treasury Department have required that a compelling data-driven case be made for amending either statute or regulation, although clearly, it is more difficult to amend the statute.

Occasionally, policy agreements crafted by the Treasury Department or the IRS and one of the statistical agencies may be used to supplement the statute and regulations. For example, the IRS-Census Bureau Criteria for the Review and Approval of Census Bureau Projects that Use Federal Tax Information (sometimes called the Criteria Agreement) has been used to delineate and clarify the process under which FTI may be accessed for new Census Bureau purposes, especially authorized research purposes at the Census Bureau as part of their Research Data Center arrangement.

CHALLENGES IN PROTECTING CONFIDENTIALITY

Protecting the confidentiality of tax data is challenging for the IRS, especially because there is no statute of limitations and because the tax code treats all FTI the same with respect to confidentiality protection. That is, to the IRS, a business name or address is as deserving of confidentiality protection as income items for a large corporation's or individual's tax return, and all must be protected in perpetuity, even after they have been anonymized as statistical tabulations for public release. Given these constraints, the resource consequences of safeguarding taxpayer confidentiality over time are nontrivial. These constraints are exacerbated by the potential for complementary disclosure, or the reidentification of taxpayer data using indirect means, for example, using data in other publicly released data to identify FTI related to a particular taxpayer. Given the ever-increasing public releases of tax and other data, the task of protecting FTI is daunting, especially over time.

CONSTRAINTS ON USING DATA

As indicated earlier, the access of FTI must be only for purposes authorized by statute, possibly supplemented with regulations and, infrequently, policy agreements. In addition, authorized recipients are subject to regular safeguards reviews in order to confirm their understanding and implementation of the many requirements covering physical and computer security, data need and use, and appropriate documentation. Other requirements include separate systems for processing or accessing FTI and background checks on individuals accessing it within facilities certified for such purposes. All these requirements are intended to preserve the confidentiality of FTI, whether maintained in its original form or commingled with data from other sources. In addition, the penalties and fines for unauthorized disclosures or inspections (also known as browsing) can be severe and are detailed in the tax code.

All of these constraints are largely driven by concerns for taxpayer confidentiality, and, in general, they seek to control or regulate the use of tax data by conceptually limiting, physically confining, and tracking such access in order to provide a documented audit trail that will withstand outside or third-party scrutiny. Implicitly, both the IRS and Congress recognize that this approach does not guarantee complete confidentiality, as the only means for such an assurance would be not to release any data at all. However, padlocking the treasure of tax data is viewed as neither a desirable nor a viable outcome, so some disclosure risk is accepted as part of the necessary balance of protection and access. The challenge is to identify acceptable risk, and the approach utilized to date is taking steps that prevent reidentification of tax data through "reasonable means." The interpretation of reasonable means includes the use of reasonably available computer technology, mathematical/statistical techniques, and a working knowledge of the related subject matter. The reasonable means standard attempts to avoid system meltdown in the use of FTI. "Reasonable means" is a technology-relative concept and thus, it may be a moving target. Nevertheless, it represents an attempt at balancing the two goals for tax data: their protection and their effective use.

It seems clear that there is probably some overall limit on tax data access, even if that limit is not precisely known. The need for this limit can be attributable to both resource costs of protection and what might be termed as the perception of a plausible quantity limit on access. To see why such a limit makes sense, consider that even large amounts of safeguarding resources cannot enable unlimited access to FTI. The reason is credibility. It is simply not credible that unlimited access would ever pass a perceptions test on confidentiality protection, especially for third-party scrutiny. That is, such an outcome would not seem plausible, as it would seem to turn the very concept of confidentiality on its head.

SOMETIMES CONFLICTING MANDATES

Statistical agencies such as the Census Bureau are mandated to use existing data systems (especially administrative records) to the maximum extent possible. Combined with the IRS statutory mandate to provide FTI only to the minimum extent needed for authorized purposes, these mandates create a tension that drives a need to negotiate the appropriate amount of FTI accessible for a given statistical task. This is not to say that the relationship between tax agency and nontax agency needs to be very adversarial. In effect, access to FTI should be treated as a scarce resource. Accordingly, the opposing mandates create an initial starting point that requires interagency cooperation in order to find a welfare-improving outcome, in which both parties find it in their interest to move to this new point. Thus, while tension from the conflicting mandates may be viewed initially as a problem, it is probably necessary to ensure the protection of taxpayer confidentiality and the provision of FTI only to the extent necessary for compelling statistical needs. Without such tension, there would probably be some bias—either too much access or too little. With this constraint, the IRS and the authorized statistical agencies are compelled to bargain hard toward an equilibrium that upholds their respective mandates, and that both sides are willing to defend. Ultimately, any interagency agreement must be documented in order to be clearly implemented and to successfully withstand outside scrutiny. Thus, the conflict in mandates provides a type of pricing mechanism for achieving a balance between supply and demand for FTI access. Forces likely to continue exerting pressure on this mechanism would include declining survey response rates, statistical processing costs, response burden, and, of course, the need to maintain voluntary tax compliance by protecting taxpayer confidentiality.

THE CENSUS BUREAU-IRS CRITERIA AGREEMENT

A 1999-2000 IRS safeguards review of the Census Bureau raised concerns over access by its research data centers (RDCs) to FTI, especially from the perspective of statutorily authorized purpose. As a result, the Census Bureau and IRS agreed to the terms of the coauthored Criteria for the Review and Approval of the Census Bureau Projects that Use Federal Tax Information, effective September 19, 2000. The Criteria Agreement outlined protocols and other requirements governing access to FTI for new uses by the Census Bureau, especially for RDC projects. The IRS review role, assigned to the Statistics of Income Division, consists of approving or concurring on the predominant Title 13, Chapter 5, purpose of proposed projects and ensuring that the minimal FTI needed would be accessed for a given proposed usage. Scientific merit remained the province of the Census Bureau and the researcher community, and, for purposes of the Criteria Agreement, outside researchers were treated as Census Bureau employees (under the special sworn status designation). The Criteria Agreement can be seen as a good outcome for the opposing agency mandates governing access to FTI for statistical purposes. As a policy agreement, it established an explicit interagency standard for authorized purpose that supplemented long-standing statutes and regulations in adapting to changing user needs. A cornerstone of the agreement was its emphasis on proposal review documentation, including explicit dual agency approvals on both the project proposals and post-project certifications. In addition, the review process it fostered implicitly enlisted the research community's active participation by forcing it to develop and maintain review capital and adding to the interagency appreciation of confidentiality needed to make this process viable. Such an outcome recognizes the limited review resources available in both the Census Bureau and IRS and was essential in order to promote a viable flow of projects. The process also made all three participants—the Census Bureau, the IRS, and the researcher community—aware of the need to work together in order to make the process demonstrably credible for purposes of potential third-party scrutiny.

DATA-SHARING PROPOSALS

For three decades, the federal statistical community has attempted to overcome certain deficiencies—particularly list frame coverage across agencies—of the decentralized data collection system by submitting a number of proposed statistical data-sharing bills to Congress. These usually died in committee after being introduced, and many required an amendment to the tax code, due to the importance of tax data in these proposals, and because FTI for business is inextricably commingled with non-FTI on the Census Bureau's business register. As a result, a number of the data-sharing proposals were accompanied by companion tax bills, known as "J bills," due to their proposed amendment of Section 6103(j). In the 1990s, two of these proposals addressed both demographic and business data and encompassed all 10 major statistical agencies. Neither was enacted.

The most recent data-sharing legislation, part B of the Confidential Information Protection and Statistical Efficiency Act (CIPSEA), focused on sharing only business data and was restricted to the three major business data statistical agencies: BEA, the Census Bureau, and the Bureau of Labor Statistics (BLS). A companion J bill accompanied CIPSEA when it was introduced in Congress in July 2002. This strategy led to the enactment of partial (nontax) data sharing when CIPSEA was signed into law in December 2002. However, the accompanying J bill attracted less legislative support, never made it to the floor of the House, and expired with that Congress.

From this experience a number of lessons have emerged, some of which CIPSEA had already absorbed, and some of which, with hindsight, might have led to some differences in both approach and content.

LESSONS LEARNED

Strong Leadership Is Needed

Strong leadership was provided for CIPSEA throughout the 2002 effort by both the Council of Economic Advisers and the Office of Management and Budget. More advocates were probably needed, especially in Congress, in order to advance both (tax and nontax) parts of the proposal once it arrived there. In fact, support by members in both the House and the Senate would seem critical in order to overcome concerns about increased sharing of identifiable tax data. In a similar vein, the support of congressional staff—especially on the tax-writing committees—needs to be enlisted, with a compelling case on why the bill is needed, including not only how government statistical operations would be improved but also how taxpayer confidentiality would remain protected.

Dispell the Myth of Access Being a Zero-Sum Activity

Part of the education effort needed would be well spent focusing on the myth in the tax community that expanded access to FTI is undesirable in general and that access cannot be expanded in one area without a commensurate reduction somewhere else. For example, the notion that increasing the number of agencies accessing business data can only be accomplished at the expense of reducing another agency's existing access to FTI must be countered. One strong argument countering this position is the evidence provided by adding both USDA-NASS and CBO to the statute as authorized recipients of FTI without reduction in access elsewhere and without observable problems in terms of weakened confidentiality. Another counterargument might be the controlled expansion in access enabled by the Census Bureau-IRS Criteria Agreement for the RDCs—now in its sixth year of implementation.

Discrete Steps May Be Better than Bold Leaps

Concerns articulated by some opponents of the 2002 CIPSEA effort include the notion that the J bill's expanded access seemed too broad because (1) it was modeled on the Census Bureau-IRS Criteria Agreement, which included access by researchers at the Census Bureau RDCs, and (2) the statutory language referenced regulations to be released in the future for purposes of authorizing access to specific items of FTI. The two major purposes of sharing FTI for Part B of CIPSEA have largely been described as establishing a common business list frame for all three agencies (BEA, the Census Bureau, and BLS) and in providing common identifier information that would enable the three agencies to exchange nontax data with each other. The potential for excessive access to tax data under the CIPSEA J bill was a concern, given that expanding item access via regulation would require only Treasury approval, not an act of Congress. This concern may have been heightened by the interest BLS has indicated in sharing limited FTI with its state partners. Moreover, both BEA and BLS might some day want to pursue access arrangements with researchers in a way similar to what the Census Bureau has done with its RDC model, which would increase the number of persons with FTI access. In short, these concerns were raised about the ability to limit, track, or control access to FTI and should be addressed by any future J bill.

One possibility for assuaging such concerns is to stipulate a limited amount of FTI in the statute itself, obviating the need for regulations. The items themselves might be limited to, say, taxpayer identification number (TIN), name, address, industry code, and one or two magnitude variables, such as employment size and income, for the purpose of stratifying a sample. Listing in the statute only the items needed for purposes of addressing the central problem (i.e., mutual list frame coverage, exchange of nontax data) might help emphasize the agencies' good faith effort to request and use only the items justified by the data-sharing rationale, so that the principle of the minimum FTI needed would be met. In addition, such a statutory limitation might help signify that these agencies did not intend to replace surveys with FTI per se, an argument sometimes raised by opponents of expanded access.

Show Some Benefit to the Treasury

When CIPSEA and the J bill were introduced to Congress in July 2002, staff from both the House Ways and Means and the Senate Finance Committees raised questions about how the legislation, especially the data sharing enabled by the J bill, would benefit the Treasury Department. CIPSEA seemed to contemplate a one-way flow of data for statistical purposes, which did not appear to include statistical tax analysis conducted by the Office of Tax Analysis in the Treasury Department and the Statistics of Income Division at the IRS. Thus, the general consensus seemed to be that Treasury would not directly benefit much by CIPSEA and the accompanying J bill.

One way of addressing this concern might be for Treasury (and other outside analysts) to benefit from the creation and release of more public-use files, including those created with synthetic data. Two problems attend this recommendation. First, the jury is still out on the utility of public-use files created with synthetic data. Second, virtually no public-use files of business data exist due to the difficulty of masking the interesting data features of concentrated industry activity at the same time that these properties are needed for analysis. A more direct way to provide Treasury with analytical benefit might be for the Census Bureau to consider Treasury and the IRS researchers for RDC access on meritorious project proposals, as long as they adhere to the same requirements as other researchers, including predominant Title 13 purpose. Preliminary discussions between the IRS and the Census Bureau so far indicate that this might have value.

One possible objection to this idea pertains to Section 7214(a)(8) of the tax code. This statute requires Treasury and IRS employees with evidence of revenue law violations to report it. The Census Bureau's concern is that this obligation might overshadow the confidentiality oaths required for special sworn status at the Census Bureau. Several factors might help mitigate this concern. For example, the statute's evidentiary standard on what constitutes information that a revenue violation has occurred is high, and it is unlikely to be uncovered during the sort of research and analysis that Treasury or the IRS might propose, especially given the limited data related to actual tax liability available at an RDC. An additional point is that any Treasury or IRS researcher would most likely be intent on statistical research. Enforcement personnel, such as auditors or tax examiners, would not be likely candidates for access, so there would be little emphasis on case-by-case compliance issues. This standard would be consistent with the engagement of some researchers at RDCs who represent commercial enterprises with a variety of clients. That is, the suspension of a non-RDC allegiance for purposes of data access is hardly unprecedented in the Census Bureau's RDC experience.

Interagency Disclosure Coordination Needed

As additional assurance to reviewers of a future J bill, it may be advisable to consider making explicit, perhaps in the narrative accompanying the bill, that the agencies authorized to share FTI would collaborate on statistical disclosure limitation methodologies. It is probably important that such coordination be given clear prominence in the J bill itself, given that the statute (Section 6103(j)(4)) requires that both direct and indirect means of reidentification be prevented with any public release of data. All FTI, including the building blocks of any list frame, such as name, address, and TIN, will probably remain subject to perpetual protection under provisions of the tax code, as discussed previously. Accordingly, legislators who understand that the statistical community has the issue of taxpayer and respondent confidentiality foremost in mind as it seeks expanded access to confidential data may be more sympathetic to a future J bill. Demonstrating such care and foresight may also assist with any future proposals that might expand data sharing beyond three agencies and encompass more than business data.

CONCLUSION

We view FTI as a national asset that can have great value in many situations faced by the statistical community. However, this asset comes with numerous constraints on its use, in particular, a strong emphasis on taxpayer confidentiality and a requirement that only the minimum amount of FTI be provided to meet authorized uses. These constraints can be productively addressed through good faith bargaining between the IRS and the statistical agencies authorized to receive FTI. We believe this bargaining process is a useful way to strike the right balance between needed access to FTI and concerns for taxpayer confidentiality. Future expansions of the statutory provisions allowing access to FTI are possible, but will take a concerted effort by the affected federal statistical agencies. Learning from past efforts can help increase the chances of success in this endeavor.

7

The Importance of Data Sharing to Consistent Macroeconomic Statistics

Dennis Fixler and J. Steven Landefeld
Bureau of Economic Analysis

The Bureau of Economic Analysis (BEA) has a unique position in the decentralized U.S. statistical system. BEA produces the national income and product accounts (NIPAs), a comprehensive and consistent double-entry set of accounts for the economy. BEA uses a myriad of data collected from public and private data sources to construct these accounts. In this role, BEA often confronts major inconsistencies in piecing these data together that are not evident from the perspective of the agencies collecting the individual pieces of the economic puzzle. BEA has been described as the canary in the mineshaft for the U.S. statistical system.

The U.S. statistical system has evolved over time in such policy agencies as the U.S. Department of Commerce, Labor, the Treasury, Agriculture, and Energy to provide data and answer questions relevant to the agencies' missions. Surveys and the legislation supporting them have evolved independently. The result is a diverse set of data using different business registers, different industry classifications for establishments, different concepts and definitions, different timing, and different collection methods.

These differences in survey frames and procedures produce significant quantitative differences in what would appear to be the same measures of economic activity. For example, employment in individual industries as reported by the Labor Department's Bureau of Labor Statistics (BLS) can differ markedly from that reported by the Department of Commerce's Census Bureau. Differences exist for wages and salaries across industries, across states, and in the aggregate (see Tables 7-2 and 7-6).

WHY IT MATTERS

The implications of these differences in estimates is illustrated in Table 7-1, which summarizes the various data sources used by BEA in constructing one of its sets of accounts. Gross domestic product (GDP) is mainly estimated using data collected by the Census Bureau, while gross domestic income (GDI) is mainly estimated using data collected by BLS, the Census Bureau, and the Statistics of Income (SOI, part of the Internal Revenue Service, IRS). In concept, GDP should equal GDI because all final expenditures should end up as income to households, business, or government. However, because of the differences in the source data used in estimating GDP and GDI, often they are not equal, and the result is the statistical discrepancy.

Such discrepancies between GDP and GDI can have large impacts on fiscal and monetary policy. During the latter half of the 1990s, a large and persistent discrepancy arose, with real GDI growing 0.6 percent faster than real GDP (1995-2000). This was important for budget planning because real trend GDP growth is used as the baseline for estimating near-term trend growth in 5-year budget forecasts made by the Office of Management and Budget (OMB) and the Congressional Budget Office. To illus-

TABLE 7-1 BEA Summary Account 1—Primary Data Sources (billions of dollars)

	Primary Data Source	2004
Income side		
Labor compensation	BLS	$6,693.4
Corporate profits & gov't enterprises	Census Bureau, SOI	973.6
Proprietors' income and rental income	Census Bureau, SOI	1023.8
Interest on assets, taxes, & misc. payments	SOI, FRB	1,531.3
Depreciation	Census Bureau	1,435.3
GROSS DOMESTIC INCOME		$11,657.5
Statistical discrepancy		76.8
GROSS DOMESTIC PRODUCT		$11,734.3
Expenditure side		
Personal consumption expenditures	Census Bureau	$8,214.3
Gross private domestic investment	Census Bureau	1,928.1
Gov't consumption exp. & gross invest.	Gov't, Census Bureau	2,215.9
Net exports of goods and services	Census Bureau, BEA	–624.0
GROSS DOMESTIC PRODUCT		$11,734.3

trate the impact, according to OMB's FY 2006 analysis of the "sensitivity of the budget to economic assumptions," a persistent understatement of real GDP growth by 1.0 percent would result in an overstatement of the projected deficit of $530 billion. Similarly, a persistent understatement of real trend GDP growth could lower the Federal Reserve's estimate of non-inflationary sustainable growth and signal the need for a tighter monetary policy than necessary.

One possible answer to the source of this discrepancy could lie in the recording of stock options, bonuses, and fringe benefits in employee compensation. While there are many sources of the difference between BLS and the Census Bureau payroll data, it is interesting that during the latter half of the 1990s, when stock options and bonuses were growing rapidly, the Census Bureau data rose at a 7.8 percent average annual rate, whereas the BLS data rose at a 7.5 percent average annual rate (1995-2000). Part of this may reflect the recording of stock options. For example, in Washington State—a state with significant stock option activity—the Census Bureau data grew nearly twice as fast (11.5 percent) as the BLS payroll data (6.2 percent) for 2000. If it turned out that stock options were under-reported in the BLS data, it would suggest that the growth rate of GDI might be even higher, thereby focusing additional effort on improving the reporting of final expenditures on services and other less-well-measured components of GDP.

Another example of the importance of BEA accuracy is illustrated by its regional data, which are used in the geographic allocation of nearly $200 billion in federal funds. These data are also used by virtually every state for its tax and planning purposes. BEA uses BLS data for these state and local estimates, which are taken from quarterly employment and unemployment tax forms.

The differences between the two sets of payroll data across states vary from the BLS set's being 4.2 percent higher in New Mexico to 9.5 lower in Alaska than what is reported by the Census Bureau. These differences could have a significant impact on the allocation of state Medicaid funds, which uses BEA per capita state personal income to determine the federal share of payments for each state. Differences in growth rates can also have an important impact on state tax projections and spending plans. For example, in New York the $1.2 billion difference in growth in wages and salaries between 2001 and 2002 between BLS and the Census Bureau series would amount to about a $173 million difference in projected income taxes.

These are but a few of the examples of the implications for government and business decision makers. In the sections below, the implications for users of estimates ranging from profits and productivity to inflation and offshoring are explored.

DATA SHARING

Over the years, numerous proposals have been put forth to resolve the problems associated with a decentralized system. One has been the creation of a central statistical office such as those that exist in Canada, Australia, and other countries. The creation of such an entity has not proved popular for various reasons. Consolidation would require extensive budget negotiation and resources to coordinate and implement a process that ensures there is little to no disruption in data production. Furthermore, the current system allows for the specialization that has arguably led to many of the innovations produced by U.S. statistical agencies.

Given these difficulties, a practical way to achieve many of the benefits of a central statistical agency without the costs is to permit the sharing of business data among the three general-purpose statistical agencies—BEA, BLS, and the Census Bureau—that produce the bulk of the nation's economic data. All three agencies have an excellent record of protecting confidential data, have access to various types of tax data, and share various types of data that could be significantly improved by data sharing.

A major step forward in allowing data sharing was the passage of the Confidential Information Protection and Statistical Efficiency Act of 2002 (CIPSEA). Section 521 stipulated that business data can be shared for statistical purposes among BEA, the Census Bureau, and BLS. At the time CIPSEA was formulated, it was understood that for data sharing to be completely operational, there would have to be some changes in Section 6103, paragraph (j), of Title 26 (Internal Revenue Code) and the accompanying regulations that govern access to federal tax information (FTI). These changes are necessitated by the facts that much of the Census Bureau information is commingled with FTI and neither BEA nor BLS has the Census Bureau level of access to use such data. Although there have been discussions concerning the formulation of a bill to submit to Congress to bring about the necessary changes in Title 26, to date no bill has been written for submission.

The absence of fully implemented data sharing especially affects BEA because it collects few data of its own and relies primarily on the Census Bureau for its data. Data sharing, however, does not just affect the ability of BEA to access Census Bureau data; the inability of BLS and the Census Bureau to share data greatly affects the quality of the data that BEA receives from both agencies. In this chapter we provide examples of how the absence of data sharing affects BEA estimates.

The limited access to business tax data has enormous effects on BEA's ability to access Census data that are commingled with tax data. The Census Bureau sample frames are constructed from IRS data, and, under current rules, name, address, and employer identification numbers are generally considered tax data. Although in principle BEA has access to corporate tax records in the SOI sample, the Census Bureau does not know the identification of those firms and so BEA has generally not been allowed access to Census records. Without going into the arcane detail, whether BEA has access to corporate Census records that are commingled with tax data is determined by the extent to which the Census Bureau claims that data are based on their own collection and not IRS records. Such a claim is generally made by the Census Bureau in the case of multiunit establishments. Thus BEA cannot access Census Bureau records from single-unit establishments. Finally, because legislation limits BEA access to corporate tax records, BEA cannot access partnership and sole proprietor Census Bureau records, which are collected from tax data–based sample frames.

The limited access to tax data also impedes BEA's use of the Census records to construct sample frames for its international surveys. The impediment is especially problematic in the services area, because many of these providers are not multiunit establishments. In a joint effort by BEA, Census, and the National Science Foundation (NSF) regarding identifying international research and development expenditures, it was discovered that there was considerable difference between Census and BEA sample frames. In this case, BEA had identified many firms that were not in the Census Bureau sample.

Below we provide some detailed illustrations of how the absence of data sharing affects BEA estimates. We also discuss how the effect on BEA estimates would affect policy decisions that are based on those estimates.

Industry Employment Differences

BLS, the Census Bureau, and SOI are the main sources of wage and salary data in the U.S. economy. Figures 7-1 and 7-2 show that the levels and growth of total payroll according to these sources are broadly consistent, but that there are significant differences in magnitudes. Below we focus on the BLS and the Census Bureau data, as those are the two main sources used by BEA.[1]

BLS prepares comprehensive wage and salary data in its Quarterly

[1]The SOI data are composed from a sample of tax returns and therefore are not as comprehensive as either BLS or the Census Bureau data. Furthermore these data are released with a lag.

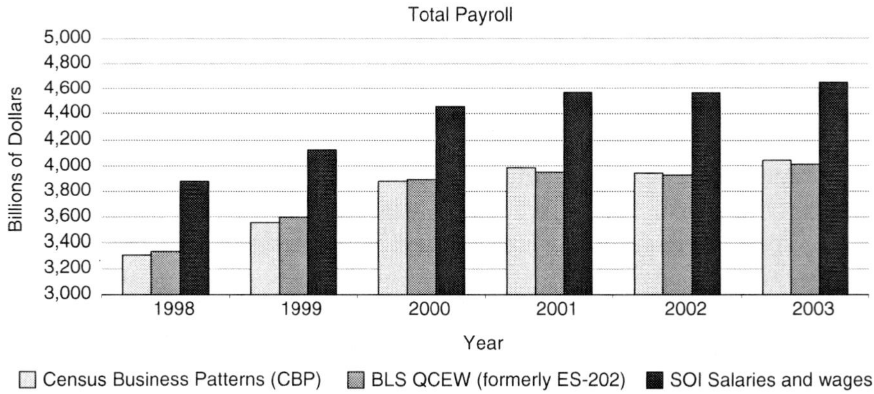

FIGURE 7-1 Payroll data comparison—the Census Bureau, BLS, and SOI levels.

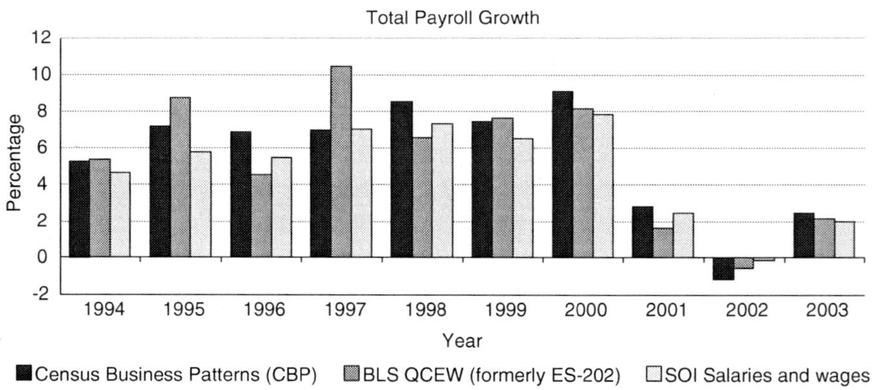

FIGURE 7-2 Payroll data comparison—the Census Bureau, BLS, and SOI growth.

Census of Employment and Wages Program (QCEW).[2] These data are widely used in BEA and are the basis for the wage and salary component of personal income. The Census Bureau also prepares payroll data as part of its Quinquennial Economic Census and Annual Survey programs. These Census data are considered to be less timely than BLS data, but in some areas, such as educational services, membership organizations, and nonprofits, they are considered to be more complete than the QCEW data.

[2]These data are commonly referred to as the ES-202 data, the former name of the program.

In addition, Census data on wages are generally recognized as providing a better industry distribution of aggregate wages, and incorporating these data into BEA estimates offers a unique opportunity to create greater consistency among the BEA industry accounts' measures of gross output, intermediate inputs, and value added. The decision, however, to choose one set of data over the other has implications for the measurement of value added in the industry accounts, which can be traced out through examining the estimates prepared as part of the annual industry accounts.

For some industries, the differences in the level of employment are significant. Table 7-2 identifies the differences in levels between BLS and the Census wage and salary data for 2002, an economic census year.[3] The primary explanations for the differences are that the Census Bureau and BLS have different sample frames of establishments and that establishments in both frames are not always classified in the same industry. The Census Bureau and BLS are currently engaged in a project that seeks to study this source of difference and explore other sources of differences. Before this project could be undertaken, approval from the IRS had to be obtained. The presentation at the workshop by James Spletzer (BLS) and Paul Hanczaryk (Census Bureau) provided details of the study (see Chapter 2 of this volume).

As shown in Table 7-2, there are many relatively large differences among industries in which estimates are available from both BLS and the Census Bureau. In the case of oil and gas extraction, the Census payroll estimate is about 50 percent lower than the BLS estimate. In addition, the Census Bureau estimate for all of manufacturing is about 15 percent—or roughly $100 billion—lower than the BLS estimate. In contrast, Census payroll estimate for management of companies and enterprises is about 63 percent—or over $70 billion—higher than the BLS estimate.[4]

Because employment and wage data are used in several places in the national accounts, we will now show how BEA estimates would be different if the Census data were used instead of the currently used BLS data for manufacturing and a few other industries in the computation of value-added. Although the current-dollar growth rate could change by as much as 2.0 percentage points (e.g., computers and electronic products), Table 7-3 shows the relative rankings for the selected industries tended to be relatively stable.

---

[3]The Census payroll data used are from the U.S. Census Bureau web site as of April 1, 2005. BLS wage data are consistent with the 2004 annual revision to the national income and product accounts and the 2004 annual revision to the annual industry accounts.

[4]This pattern may suggest a different classification treatment of head company offices by the Census Bureau and BLS.

TABLE 7-2 Differences in BLS Wages and the Census Bureau Payroll by NAICS Industry, 2002 (in thousands of dollars)

1997 NAICS Codes	Industry Name	BEA Wages[a]
	All Industries	4,968,131,000
	Private industries	4,119,730,000
11	Agriculture, forestry, fishing, and hunting	31,815,000
111, 112	Crop and animal production ("Farms")	17,685,000
113, 114, 115	Forestry, fishing, and related activities	14,130,000
21	Mining	30,788,000
211	Oil and gas extraction	11,455,000
212	Mining, except oil and gas	10,470,000
213	Support activities for mining	8,863,000
22	Utilities	40,094,000
23	Construction	272,418,000
31, 32, 33	Manufacturing	675,523,000
33, 321, 327	Durable goods	441,182,000
321	Wood product manufacturing	17,585,000
327	Nonmetallic mineral product manufacturing	20,674,000
331	Primary metal manufacturing	23,209,000
332	Fabricated metal product manufacturing	59,742,000
333	Machinery manufacturing	57,050,000
334	Computer and electronic product manufacturing	98,359,000
335	Electrical equipment and appliance manufacturing	20,630,000
3361, 3362, 3363	Motor vehicle, body, trailer, and parts manufacturing	58,705,000
3364, 3365, 3366, 3369	Other transportation equipment manufacturing	38,954,000
337	Furniture and related product manufacturing	18,232,000
339	Miscellaneous manufacturing	28,042,000
31, 32 (excluding 321 and 327)	Nondurable goods	234,341,000
311 ,312	Food product manufacturing	60,356,000
313, 314	Textile and textile product mills	14,525,000
315,316	Apparel manufacturing	10,751,000
322	Paper manufacturing	25,611,000
323	Printing and related support activities	27,061,000
324	Petroleum and coal products manufacturing	7,632,000
325	Chemical manufacturing	57,293,000
326	Plastics and rubber products manufacturing	31,112,000
42	Wholesale trade	280,745,000
44, 45	Retail trade	360,341,000
48, 49	Transportation and warehousing, excluding postal service	162,206,000
481	Air transportation	30,550,000
482	Rail transportation	11,824,000
483	Water transportation	2,888,000
484	Truck transportation	47,917,000

BLS Wages[b]	Census Payroll[c]	Percent Difference[d] BEA and Census	Percent Difference[d] BLS and Census	BLS and Census Difference[e]
		—	—	
3,923,090,541		—	—	
24,146,183		—	—	
15,862,753		—	—	
8,283,429		—	—	
30,557,227		—	—	
11,269,829	5,564,811	–51.4	–50.6	–5,705,018
10,321,353	8,987,397	–14.2	–12.9	–1,333,956
8,966,044	6,707,242	–24.3	–25.2	–2,258,802
39,895,551	43,493,804	8.5	9.0	3,598,253
260,841,814	254,000,182	–6.8	–2.6	
670,676,772	573,401,510	–15.1	–14.5	–97,275,262
437,547,486	370,407,941	–16.0	–15.3	–67,139,545
16,952,331	15,909,908	–9.5	–6.1	–1,042,423
20,543,618	17,933,376	–13.3	–12.7	–2,610,242
23,246,080	21,508,667	–7.3	–7.5	–1,737,413
59,352,280	57,361,374	–4.0	–3.4	–1,990,906
56,689,509	49,470,768	–13.3	–12.7	–7,218,741
98,045,569	64,314,150	–34.6	–34.4	–33,731,419
20,479,516	17,957,015	–13.0	–12.3	–2,522,501
58,579,129	50,331,680	–14.3	–14.1	–8,247,449
38,446,534	31,231,174	–19.8	–18.8	–7,215,360
18,107,133	17,364,837	–4.8	–4.1	–742,296
27,105,787	27,024,992	–3.6	–0.3	–80,795
233,129,286	202,993,569	–13.4	–12.9	–30,135,717
59,649,421	52,334,562	–13.3	–12.3	–7,314,859
14,501,506	12,333,814	–15.1	–14.9	–2,167,692
10,360,588	8,567,969	–20.3	–17.3	–1,792,619
25,744,232	21,336,257	–16.7	–17.1	–4,407,975
26,457,610	25,738,613	–4.9	–2.7	–718,997
7,891,082	6,202,508	–18.7	–21.4	–1,688,574
57,322,150	44,032,801	–23.1	–23.2	–13,289,349
31,202,697	32,447,045	4.3	4.0	1,244,348
276,607,852	249,986,560	–11.0	–9.6	–26,621,292
348,909,029	296,215,722	–17.8	–15.1	–52,693,307
146,810,674		—	—	–146,810,674
30,180,386		—	—	–30,180,386
10,869		—	—	–10,869
2,793,556	3,031,880	5.0	8.5	238,324
46,824,531	47,833,730	–0.2	2.2	1,009,199

Continued

TABLE 7-2 Continued

1997 NAICS Codes	Industry Name	BEA Wages[a]
485	Transit and ground passenger transportation	8,996,000
486	Pipeline transportation	3,272,000
487, 488, 492	Other transportation and support activities	39,802,000
493	Warehousing and storage	16,957,000
51	Information	189,736,000
511	Publishing including software	58,394,000
512	Motion picture and sound recording industries	18,258,000
513	Broadcasting and telecommunications	84,838,000
514	Information and data processing services	28,246,000
52	Finance and insurance	370,088,000
521, 522	Federal Reserve banks, credit intermediation and related services	132,010,000
523	Securities, commodity contracts, investments	112,344,000
524	Insurance carriers and related activities	119,830,000
525	Funds, trusts, and other financial vehicles	5,904,000
53	Real estate, rental, and leasing	71,785,000
531	Real estate	51,015,000
532,533	Rental and leasing services and lessors of intangible assets	20,770,000
54	Professional and technical services	415,422,000
5411	Legal services	80,297,000
5415	Computer systems design and related services	84,251,000
5412-5414, 5416-5419	Other professional, scientific and technical services	250,874,000
55	Management of companies and enterprises	117,147,000
56	Administrative and waste services	193,525,000
561	Administrative and support services	180,230,000
562	Waste management and remediation services	13,295,000
61	Educational services	74,446,000
62	Health care and social assistance	472,214,000
621	Ambulatory health care services	209,724,000
622, 623	Hospitals and nursing and residential care facilities	217,119,000
624	Social assistance	45,371,000
71	Arts, entertainment, and recreation	51,526,000
711, 712	Performing arts, museums, and related activities	24,724,000
713	Amusements, gambling, and recreation	26,802,000
72	Accommodation and food services	153,922,000
721	Accommodation	40,764,000
722	Food services and drinking places	113,158,000

BLS Wages[b]	Census Payroll[c]	Percent Difference[d] BEA and Census	Percent Difference[d] BLS and Census	BLS and Census Difference[e]
7,787,298	7,575,497	–15.8	–2.7	–211,801
3,277,932	3,082,558	–5.8	–6.0	–195,374
39,101,445	34,489,516	–13.3	–11.8	–4,611,929
16,834,658	18,689,122	10.2	11.0	1,854,464
188,758,526		—	—	–188,758,526
58,307,089	64,712,028	10.8	11.0	6,404,939
17,879,785	12,516,040	–31.4	–30.0	–5,363,745
84,664,461	88,624,463	4.5	4.7	3,960,002
27,907,191	27,686,444	–2.0	–0.8	–220,747
356,371,058		—	—	–356,371,058
131,188,066	124,076,870	–6.0	–5.4	–7,111,196
108,325,327	101,285,387	–9.8	–6.5	–7,039,940
110,965,984	120,683,183	0.7	8.8	9,717,199
5,891,681		—	—	–5,891,681
68,801,129		—	—	–68,801,129
48,110,832	41,911,444	–17.8	–12.9	–6,199,388
20,690,296	18,706,319	–9.9	–9.6	–1,983,977
390,450,138		—	—	–390,450,138
69,875,728	69,939,404	–12.9	0.1	63,676
83,897,952	72,168,495	–14.3	–14.0	–11,729,457
236,676,458		—	—	–236,676,458
117,462,176	190,807,531	62.9	62.4	73,345,355
191,825,310		—	—	–191,825,310
178,563,429	195,425,035	8.4	9.4	16,861,606
13,261,881	12,178,484	–8.4	–8.2	–1,083,397
64,700,545		—	—	–64,700,545
456,030,369		—	—	–456,030,369
204,320,753	203,716,200	–2.9	–0.3	–604,553
215,390,850	212,480,514	–2.1	–1.4	–2,910,336
36,318,766	36,090,970	–20.5	–0.6	–227,796
47,050,671		—	—	–47,050,671
24,652,961	24,057,801	–2.7	–2.4	–595,160
22,397,710	21,069,716	–21.4	–5.9	–1,327,994
142,208,429		—	—	–142,208,429
36,805,629	34,874,261	–14.4	–5.2	–1,931,368
105,402,801	92,632,794	–18.1	–12.1	–12,770,007

Continued

TABLE 7-2 Continued

1997 NAICS Codes	Industry Name	BEA Wages[a]
81	Other services, except government	155,989,000
	Government	848,401,000
	Federal civilian	141,631,000
	Federal military	59,529,000
	State and local	647,241,000

NOTE: Census payroll data are from the U.S. Census Bureau web site as of April 1, 2005. BLS wage data are consistent with the 2004 annual revision of the national income and product accounts and the 2004 annual revision of the annual industry accounts.

[a]Wage and salary disbursements consists of the monetary remuneration of employees, including corporate officers salaries and bonuses, commissions, pay-in-kind, incentive payments, and tips. It reflects the amount of payments disbursed, but not necessarily earned during the year. Wage and salary disbursements are measured before deductions, such as social security contributions and union dues. In recent years, stock options have become a point of discussion. Personal income includes stock options of nonqualified plans at the time that they have been exercised by the individual. Stock options are reported in wage and salary disbursements. The value that is included in wages is the difference between the exercise price and the price that the stock options were granted.

Estimates of value-added in the annual industry accounts are prepared in a two-part process. First, three-digit North American Industry Classification System (NAICS) industry estimates are controlled to the national income and product accounts for compensation of employees and "taxes on production and imports less subsidies," and initial estimates of gross operating surplus are extrapolated from the most recent set of "balanced" gross operating surplus estimates.[5] Second, these three-digit NAICS industry controls are distributed to greater industry detail in the annual input-output tables through a two-step process. Detailed industry levels are extrapolated using QCEW data for compensation, and "taxes less subsidies" and gross operating surplus are extrapolated using detailed gross output estimates. Then all three components are scaled back to the three-digit controls. Extrapolation of industry detail for compensation could result in differences in shares depending on the choice to use

[5]The recently adopted integration methodology for the annual industry accounts allows for intermediate inputs and gross operating surplus to adjust during the iterative row-and-column balancing procedure. For a discussion of the integration methodology, see Moyet et al. (2004)

BLS Wages[b]	Census Payroll[c]	Percent Difference[d] BEA and Census	BLS and Census	BLS and Census Difference[e]
100,987,088		—	—	−100,987,088
		—	—	0
		—	—	0
		—	—	0
		—	—	0

[b]Reported quarterly total wages are the wages paid by Unemployment Insurance covered employers during the calendar quarter, regardless of when the services were performed. Reported total annual wages are the sum of the total wages reported for the corresponding quarters.

[c]Payroll includes all forms of compensation, such as salaries, wages, commissions, dismissal pay, bonuses, vacation allowances, sick-leave pay, and employee contributions, to qualified pension plans paid during the year to all employees. For corporations, payroll includes amounts paid to officers and executives; for unincorporated businesses, it does not include profit or other compensation of proprietors or partners. Payroll is reported before deductions for social security, income tax, insurance, union dues, etc. This definition of payroll is the same as that used by the IRS on Form 941.

[d]Computed as Census Payroll less BLS wages, divided by BLS wages.

[e]Computed as Census Payroll less BLS wages.

QCEW or the Census data. Again, these differences would be the largest for detailed industries in which the magnitude of the difference is the greatest.

The employment differences also have an impact on the computation of chain-type quantity indexes, real value-added by industry, and contributions to real growth. To measure the impact, BEA's double-deflation method for preparing real value-added for the industry accounts was simulated to incorporate different nominal value-added levels, and then the resulting impact on the real value-added estimates was examined. For this exercise, three-digit NAICS industry estimates for value-added were allowed to increase (decrease) by the difference in wage data between BLS and the Census Bureau, and new levels of nominal intermediate inputs were computed as the difference between published gross output by industry and the simulated value-added by industry. Next, the new intermediate input levels were deflated by the published price indexes for intermediate inputs to produce real intermediate inputs. Finally, real value-added by industry was computed as the difference between published real gross output by industry and real intermediate inputs by industry.

The choice of wage data affects the resulting change in real value-

TABLE 7-3 Differences in Current-Dollar Value-Added Growth by Manufacturing Industries and for ICT-Producing Industries Combined, 2002

1997 NAICS Code	Industry	2002 Published	2002 Simulated	Published Rank	Simulated Rank
31, 32, 33	**Manufacturing**	**0.4**	**0.7**		
33, 321, 327	**Durable goods**	**–0.9**	**–0.4**		
321	Wood product manufacturing	–4.2	–4.4	15	16
327	Nonmetallic mineral product manufacturing	–3.5	–3.7	14	15
331	Primary metal manufacturing	1.2	1.7	8	8
332	Fabricated metal product manufacturing	–2.4	–2.3	13	13
333	Machinery manufacturing	–5.4	–5.4	17	17
334	Computer and electronic product manufacturing	–4.7	–2.7	16	14
335	Electrical equipment and appliance manufacturing	–6.2	–6.3	18	18
3361, 3362, 3363	Motor vehicle, body, trailer, and parts manufacturing	10.0	10.2	1	1
3364, 3365, 3366, 3369	Other transportation equipment manufacturing	1.1	0.8	9	9
337	Furniture and related product manufacturing	–0.7	–0.6	10	10
339	Miscellaneous manufacturing	3.9	3.9	5	5
31, 32 (excluding 321 and 327)	**Nondurable goods**	**2.3**	**2.1**		
311, 312	Food product manufacturing	3.2	3.1	7	7
313, 314	Textile and textile product mills	–1.8	–1.6	11	11
315, 316	Apparel manufacturing	8.6	10.2	2	1
322	Paper manufacturing	3.8	3.9	6	5
323	Printing and related support activities	–1.8	–1.8	11	12
324	Petroleum and coal products manufacturing	–23.1	–24.9	19	19
325	Chemical manufacturing	6.2	6.0	3	3
326	Plastics and rubber products manufacturing	4.5	4.4	4	4
	Addenda:				
	ICT-producing industries[a]	–2.3	–1.4		

NOTE: The 3-digit NAICS industry estimates for value added were allowed to increase (decrease) by the difference in wage data between BLS and the Census Bureau. Estimates were simulated based on data published as part of the 2004 annual revision to the annual industry accounts.

[a]Consists of computer and electronic products; publishing industries (includes software); information and data processing services; and computer systems design and related services.

TABLE 7-4 Differences in Real Value-Added Growth by Manufacturing Industries and for ICT-Producing Industries Combined, 2002

		2002			
1997 NAICS Code	Industry	Simulated	Published	Published Rank	Simulated Rank
31, 32, 33	**Manufacturing**	**2.3**	**2.8**		
33, 321, 327	**Durable goods**	**1.3**	**2.0**		
321	Wood product manufacturing	–3.2	–3.4	16	16
327	Nonmetallic mineral product manufacturing	–5.2	–5.4	18	18
331	Primary metal manufacturing	1.0	1.5	9	8
332	Fabricated metal product manufacturing	–2.9	–2.9	14	14
333	Machinery manufacturing	–5.9	–6.0	19	19
334	Computer and electronic product manufacturing	7.4	15.6	4	3
335	Electrical equipment and appliance manufacturing	–4.4	–4.2	17	17
3361, 3362, 3363	Motor vehicle, body, trailer, and parts manufacturing	15.0	16.0	2	2
3364, 3365, 3366, 3369	Other transportation equipment manufacturing	–1.1	–1.5	11	11
337	Furniture and related product manufacturing	–3.1	–3.2	15	15
339	Miscellaneous manufacturing	1.1	1.1	8	10
31, 32 (excluding 321 and 327)	**Nondurable goods**	**3.7**	**3.8**		
311, 312	Food product manufacturing	–1.9	–2.5	12	13
313, 314	Textile and textile product mills	0.7	1.2	10	9
315, 316	Apparel manufacturing	10.1	11.8	3	4
322	Paper manufacturing	5.2	5.2	6	6
323	Printing and related support activities	–2.3	–2.3	13	12
324	Petroleum and coal products manufacturing	31.6	34.6	1	1
325	Chemical manufacturing	6.6	7.1	5	5
326	Plastics and rubber products manufacturing	3.5	3.4	7	7
	Addenda:				
	ICT-producing industries[a]	2.0	3.6		

NOTE: BEA's double-deflation methodology for preparing real value added for the industry accounts was simulated to incorporate different nominal value-added levels and then the attending impact on the real value added was examined. For this exercise, three-digit NAICS industry estimates for value added were allowed to increase (decrease) by the difference in wage data between BLS and the Census Bureau, and new levels of nominal intermediate inputs were computed as the difference between published gross output by industry and the simulated value added by industry. Estimates were simulated based on data published as part of the 2004 annual revision to the annual industry accounts.

[a]Consists of computer and electronic products; publishing industries (includes software); information and data processing services; and computer systems design and related services.

added for an industry. In general, the impact is greatest for industries with the largest differences in BLS-Census payroll levels. For example, real value-added for computer and electronic products would double that of the published estimates (15.6 versus 7.4 percent) if the 2002 Census rather than BLS payroll data had been used (see Table 7-4). This impact on the real value-added growth for computers would have resulted in a different story for the recovery of manufacturing in 2002: the published estimates show real growth for manufacturing increasing 2.3 percent; the simulated estimates result in a 2.8 percent real growth. In addition, the real value-added growth for the ICT-related industries[6] would have increased by 3.6 percent in 2002, or 1.6 percentage points over the published estimate, if BEA used the Census Bureau payroll data. The shifts in the industry distribution of real value-added growth shown in Table 7-4 also affect the estimates for contributions to growth and productivity.

In addition, the changing labor and capital shares for an industry that result from the changes in the computation in real value-added would in turn produce changes in the weighting of each industry's value-added price index. An industry's value-added price index represents the prices of its primary factors of production. Thus, the industry's price index and the current-dollar components of its value-added can be used to assess the contribution of each component to the value-added price index. To illustrate, suppose that the Census payroll data were used in place of the BLS wages in preparing a measure of current-dollar value-added for the oil and gas extraction industry. Compensation of employees would fall by the amount of the difference in wage data or by approximately 50 percent (see Table 7-2), thereby reducing current-dollar value-added by the difference between the two wage measures. Thus, compensation of employees as a percentage of current-dollar value-added for the oil and gas extraction industry would fall from approximately 19 to 13 percent, and gross operating surplus would rise from approximately 66 to 72 percent, thereby increasing the cost of capital.[7]

In the annual industry accounts, value-added unit costs are computed by dividing current-dollar value-added and its components by real (chained-dollar) value-added. The resulting quotients provide the value-

---

[6]Consists of computer and electronic products; publishing industries (includes software); information and data processing services; and computer systems design and related services.

[7]Shares of current-dollar value-added were computed from the data in Table 7-2 and from published data on current-dollar value-added by industry found in the GDP-by-industry accounts. The published current-dollar data are available on the BEA web site, at http://www.bea.gov/bea/dn2/gdpbyind_data.htm. Choose the GDPbyInd_VA_NAICS.xls file under the header "1998-2004 NAICS data."

added chain-type price indexes and the component price indexes. That is, unit cost measures provide estimates of a component's share of the value-added price index for an industry. The changes in the share for unit labor costs, reflected by the change in compensation of employees and for unit capital costs and embedded in the change in gross operating surplus, will have an impact on the value-added unit costs as well.[8] The reason is that value-added unit costs attribute changes in the value-added unit prices to the components of value-added in proportion to the component's share of current-dollar value-added. As a result, year-to-year changes in component shares of current-dollar value-added result in changes in the contributions of the cost components to value-added prices even if the prices do not change.

Table 7-5 presents ratios of gross output, as measured by BEA, to the different measures of employment. The data for gross output come from the preliminary 2002 benchmark input-output accounts instead of the GDP-by-industry measures used above. Also, the Census employment data are of a different vintage: data are as of October instead of the April data used in the other tables. In general, the absolute value of the difference between the NIPA ratio and the Census ratio exceeds that of the NIPA ratio and the BLS ratio. In addition, there are several industries for which there is a sign difference. If these ratios are viewed as indicators of industry productivity, one would tell different stories for many industries. However, comparisons of the rankings of industries by the ratio reveal that there are no substantive differences between them; that is, the rankings differ by several places, but the top and the bottom of the rankings contain similar industries.

Analysis of Impact on Regional Estimates

The choice of wage data affects the analysis of state economic activity. At the state level the range of differences in total private wages and salaries in 2003 vary from BLS being 4.2 percent higher in New Mexico to 9.5 percent lower in Alaska than what is reported by the Census Bureau in its County Business Patterns (CBP) data (see Table 7-6). Although the U.S. level of BLS wages is lower by only 0.6 percent, or $25.1 billion, New York's BLS data is lower than the Census data by 2.0 percent, or $6.7 billion, and the combination of Connecticut, New Jersey, and New York is lower than the Census by $13.0 billion.

The differences between the two programs have implications for

[8]Gross operating surplus in the annual industry accounts reflects a measure of capital inputs and net profits. For more information, see Strassner et al. (2005).

TABLE 7-5 Employment Data and Ratios from the BEA, Census, and BLS, 2002

2002 NAICS Code	Industry Name	BEA Gross Output (billions of dollars)[a]
11	Agriculture, forestry, fishing, and hunting	270.6
111-112	Farms	220.4
113-115	Forestry, fishing, and related activities	50.1
21	Mining	179.1
211	Oil and gas extraction	103.7
212	Mining, except oil and gas	48.0
213	Support activities for mining	27.4
22	Utilities	314.7
23	Construction	909.2
31-33	Manufacturing	3,839.2
321, 327, 331-335, 3361-3366, 3369, 337, 339	Durable goods	2,080.5
321	Wood products	87.5
327	Nonmetallic mineral products	93.0
331	Primary metals	138.2
332	Fabricated metal products	243.4
333	Machinery	241.2
334	Computer and electronic products	353.2
335	Electrical equipment, appliances, and components	100.5
3361-3363	Motor vehicles, bodies and trailers, and parts	463.6
3364-3366, 3369	Other transportation equipment	162.8
337	Furniture and related products	74.2
339	Miscellaneous manufacturing	122.7
311-316, 322-326	Nondurable goods	1,758.7
311-312	Food and beverage and tobacco products	562.3
313-314	Textile mills and textile product mills	75.2
315-316	Apparel and leather and allied products	47.7
322	Paper products	151.8
323	Printing and related support activities	94.7
324	Petroleum and coal products	212.4
325	Chemical products	444.8
326	Plastics and rubber products	169.8
42	Wholesale trade	866.6
44-45	Retail trade	1,046.0
48-49	Transportation and warehousing	603.6
481	Air transportation	98.1
482	Rail transportation	45.5
483	Water transportation	23.7
484	Truck transportation	202.3
485	Transit and ground passenger transportation	31.6
486	Pipeline transportation	28.0
487-488, 492	Other transportation and support activities	129.9
493	Warehousing and storage	44.5

NIPA Employment[b]	Census Employment[c]	BLS Employment[d]	NIPA Ratio	Census Ratio	BLS Ratio
Thousands of Employees			Output per Employee in Dollars		
1,564	N/A	N/A	173,018	N/A	N/A
870	N/A	N/A	253,333	N/A	N/A
694	N/A	N/A	72,190	N/A	N/A
511	485	507	350,489	369,548	353,411
124	105	121	836,290	991,595	853,547
213	197	210	225,352	243,214	228,932
175	183	176	156,571	149,964	156,025
594	663	591	529,798	474,629	532,493
6,978	7,374	6,694	130,295	123,306	135,822
15,349	14,693	15,218	250,127	261,299	252,280
9,528	9,052	9,454	218,356	229,830	220,064
574	540	554	152,439	162,102	157,953
520	484	517	178,846	192,311	179,816
506	495	506	273,123	279,422	273,160
1,552	1,573	1,545	156,830	154,736	157,569
1,229	1,164	1,221	196,257	207,158	197,561
1,500	1,261	1,497	235,467	280,024	235,913
498	492	496	201,807	204,486	202,626
1,153	1,088	1,152	402,082	425,957	402,594
679	608	678	239,764	267,911	240,188
607	596	604	122,241	124,467	122,948
708	752	685	173,305	163,151	179,020
5,822	5,640	5,764	302,078	311,803	305,123
1,760	1,666	1,743	319,489	337,432	322,570
487	449	484	154,415	167,639	155,416
419	387	404	113,843	123,156	118,037
542	488	541	280,074	311,126	280,597
724	719	706	130,801	131,795	134,088
117	103	118	1,815,385	2,062,617	1,798,598
928	846	923	479,310	525,512	481,711
845	982	844	200,947	172,862	201,176
5,711	5,865	5,617	151,742	147,749	154,283
15,500	14,648	15,012	67,484	71,411	69,677
4,265	N/A	3,989	141,524	N/A	151,329
562	N/A	561	174,555	N/A	174,796
194	N/A	N/A	234,536	N/A	N/A
54	66	53	438,889	358,260	445,757
1,367	1,435	1,337	147,988	140,955	151,294
403	398	372	78,412	79,321	84,946
42	37	42	666,667	761,076	660,673
1,127	1,050	1,108	115,262	123,756	117,229
516	566	514	86,240	78,687	86,524

Continued

TABLE 7-5 Continued

2002 NAICS Code	Industry Name	BEA Gross Output (billions of dollars)[a]
51	Information	956.6
511	Publishing industries (includes software)	251.3
512	Motion picture and sound recording industries	81.6
515-517	Broadcasting and telecommunications	525.6
518-519	Information and data processing services	98.2
52-53[e]	Finance, insurance, real estate, rental, and leasing	3,358.0
54-56	Professional and business services	1,838.4
54	Professional, scientific, and technical services	1,052.1
5411	Legal services	215.7
5415	Computer systems design and related services	171.5
5412-5414, 5416-5419	Miscellaneous professional, scientific, and technical services	664.9
55	Management of companies and enterprises	297.8
56	Administrative and waste management services	488.5
561	Administrative and support services	435.0
562	Waste management and remediation services	53.5
61-62	Educational services, health care, and social assistance	1,310.8
61	Educational services	152.9
62	Health care and social assistance	1,157.9
621	Ambulatory health care services	526.8
622-623	Hospitals and nursing and residential care facilities	523.4
624	Social assistance	107.7
71-72	Arts, entertainment, recreation, accommodation, and food services	709.3
71	Arts, entertainment, and recreation	175.7
711-712	Performing arts, spectator sports, museums, and related activities	82.3
713	Amusements, gambling, and recreation industries	93.4
72	Accommodation and food services	533.6
721	Accommodation	143.7
722	Food services and drinking places	389.9
81	Other services, except government	452.7

[a]Gross output data were obtained from the BEa web site at the following address: http://www.bea.gov/bea/dn2/i-o_benchmark_2002.htm (October 7, 2005).

[b]Data were obtained from the NIPA tables at the BEA web site at the following address: http://www.bea.gov/bea/dn/nipaweb/TableView.asp#Mid (October 7, 2005).

[c]Data were obtained from the U.S. Census Bureau web site at the following address: http://www.census.gov/econ/census02/data/us/US000.HTM (October 7, 2005).

[d]Data were obtained from the BLS' Quarterly Census of Employment and Wages (October 12, 2005).

NIPA Employment[b]	Census Employment[c]	BLS Employment[d]	NIPA Ratio	Census Ratio	BLS Ratio
Thousands of Employees			Output per Employee in Dollars		
3,381	3,736	3,359	282,934	256,045	284,821
998	1,090	961	251,804	230,638	261,548
393	303	385	207,634	269,176	211,980
1,507	1,772	1,534	348,772	296,694	342,739
484	572	479	202,893	171,737	204,877
7,999	8,546	7,724	419,803	392,955	434,753
16,418	18,649	15,926	111,975	98,578	115,434
7,103	7,302	6,662	148,121	144,084	157,915
1,279	1,170	1,116	168,647	184,422	193,209
1,148	1,107	1,144	149,390	154,874	149,851
4,676	5,025	4,402	142,194	132,318	151,059
1,685	2,605	1,682	176,736	114,306	177,041
7,630	8,742	7,581	64,024	55,881	64,434
7,311	8,410	7,264	59,499	51,724	59,887
319	332	318	167,712	161,232	168,376
16,752	N/A	15,353	78,247	N/A	85,375
2,709	N/A	1,961	56,442	N/A	77,985
14,043	14,010	13,393	82,454	82,651	86,457
4,758	4,938	4,630	110,719	106,681	113,790
6,893	6,965[f]	6,846	75,932	75,141	76,448
2,392	2,106	1,917	45,025	51,142	56,189
12,255	11,970	12,000	57,878	59,258	59,109
1,910	1,849	1,802	91,990	95,041	97,528
495	546	494	166,263	150,835	166,705
1,415	1,303	1,308	66,007	71,678	71,415
10,345	10,121	10,198	51,581	52,722	52,322
1,793	1,813	1,769	80,145	79,247	81,242
8,551	8,308	8,430	45,597	46,933	46,254
6,859	N/A	4,253	66,001	N/A	106,443

[e]Not all of the industry entitled "Funds, trusts, and other financial vehicles (525)" is included. The gross output for finance, insurance, and real estate includes the imputation for owner-occupied dwellings which causes higher output to employment ratios.

[f]The Census Bureau data for hospitals for 2002 include government and nongovernment hospitals. To make data comparable, government hospitals were not included.

TABLE 7-6 Census Bureau (CBP) Payroll and BLS (QCEW) Private Wage Comparison (billions of dollars, unless otherwise noted)

	CBP Annual Payroll[a]			Yr-Yr % Chg	
				2001-2002	2002-2003
	2001	2002	2003		
Alabama	45.2	45.5	47.1	0.7	3.5
Alaska	8.3	8.4	8.7	1.2	3.0
Arizona	60.0	61.1	64.4	1.8	5.3
Arkansas	25.8	25.9	27.0	0.5	4.3
California	521.8	510.8	520.6	–2.1	1.9
Colorado	71.5	67.8	67.9	–5.2	0.2
Connecticut	68.9	68.5	69.7	–0.6	1.8
Delaware	15.0	14.7	15.1	–2.0	2.5
D.C.	20.8	21.4	22.5	2.7	5.1
Florida	189.6	192.9	202.4	1.7	4.9
Georgia	115.9	113.8	116.3	–1.9	2.2
Hawaii	12.7	13.4	14.1	5.3	5.9
Idaho	12.4	12.6	13.1	1.7	4.3
Illinois	204.3	197.8	201.0	–3.2	1.6
Indiana	79.3	79.4	81.4	0.1	2.6
Iowa	34.5	34.8	36.0	1.1	3.4
Kansas	33.3	33.2	34.0	–0.5	2.7
Kentucky	42.6	42.5	43.8	–0.1	2.9
Louisiana	45.2	45.6	47.1	1.0	3.3
Maine	14.2	14.4	14.8	1.2	2.9
Maryland	74.2	75.0	78.9	1.1	5.1
Massachusetts	134.7	127.9	127.1	–5.0	–0.6
Michigan	142.9	142.4	144.0	–0.4	1.1
Minnesota	84.9	84.5	87.3	–0.4	3.3
Mississippi	22.7	22.8	23.6	0.2	3.8
Missouri	74.4	74.1	75.6	–0.4	2.0
Montana	7.2	7.4	7.7	2.8	4.0
Nebraska	20.8	21.7	23.1	4.4	6.2
Nevada	27.5	29.3	31.3	6.4	6.9
New Hampshire	18.5	18.7	18.8	0.9	1.0
New Jersey	154.2	152.4	154.5	–1.2	1.4
New Mexico	14.8	15.1	15.8	1.9	4.9
New York	343.5	329.8	332.6	–4.0	0.9
North Carolina	103.0	101.8	104.6	–1.2	2.7
North Dakota	6.4	6.6	6.9	1.8	5.3
Ohio	156.9	154.8	157.5	–1.3	1.7
Oklahoma	33.4	33.6	33.6	0.5	0.0
Oregon	44.1	43.5	44.3	–1.3	1.9
Pennsylvania	169.9	169.2	174.5	–0.4	3.1
Rhode Island	13.2	13.5	14.5	3.0	7.3
South Carolina	43.8	44.0	44.6	0.4	1.4
South Dakota	7.6	7.8	8.0	1.8	2.7
Tennessee	70.7	71.3	73.2	0.8	2.7
Texas	282.3	277.8	281.6	–1.6	1.4
Utah	26.1	26.2	26.8	0.4	2.3

QCEW Private Wages[b]			Yr-Yr % Chg	
2001	2002	2003	2002	2003
44.9	45.6	46.8	1.6	2.8
7.3	7.6	7.9	3.8	3.3
62.7	63.1	65.8	0.6	4.3
25.5	25.9	26.5	1.9	2.2
518.6	508.2	521.1	–2.0	2.5
71.6	68.9	69.1	–3.7	0.3
68.3	66.8	68.0	–2.2	1.9
13.7	13.7	14.2	0.5	3.5
22.1	22.7	23.6	2.9	4.0
191.0	195.7	204.5	2.4	4.5
116.3	114.8	116.4	–1.2	1.3
13.3	13.8	14.5	3.4	5.5
12.8	12.9	13.2	0.9	2.4
199.2	196.8	198.3	–1.2	0.8
78.9	79.6	81.1	0.9	1.8
34.1	34.6	35.6	1.4	2.9
33.0	33.1	33.3	0.2	0.9
43.3	43.8	45.1	1.1	2.8
44.2	44.7	45.7	1.1	2.3
14.1	14.4	14.9	2.2	3.3
73.4	74.9	77.7	2.1	3.7
130.4	126.4	127.6	–3.0	0.9
144.0	143.0	145.3	–0.7	1.6
82.4	83.1	85.4	0.8	2.8
22.6	23.0	23.5	1.7	1.9
73.2	73.7	74.7	0.7	1.4
7.4	7.7	8.1	4.1	4.7
20.6	21.0	21.7	2.0	3.1
29.7	30.3	32.6	2.2	7.5
19.1	19.1	19.6	0.0	2.8
146.0	147.0	150.0	0.7	2.0
15.6	15.9	16.5	2.5	3.6
334.6	322.1	325.9	–3.7	1.2
101.9	101.8	103.1	–0.1	1.2
6.3	6.5	6.8	2.8	4.9
154.1	154.2	156.6	0.1	1.5
32.3	32.2	32.8	–0.3	1.7
44.0	43.7	44.5	–0.8	1.9
167.5	169.0	173.4	0.9	2.6
13.0	13.4	14.2	3.0	5.9
42.5	42.9	44.0	0.8	2.5
7.5	7.6	7.9	2.4	3.5
70.5	71.7	73.9	1.7	2.9
286.6	281.7	284.1	–1.7	0.8
25.9	25.7	26.2	–0.6	1.6

Continued

TABLE 7-6 Continued

	CBP Annual Payroll[a]			Yr-Yr % Chg	
	2001	2002	2003	2001-2002	2002-2003
Vermont	7.3	7.4	7.7	2.1	3.0
Virginia	102.5	101.7	106.1	–0.8	4.3
Washington	86.5	83.1	90.6	–3.9	9.0
West Virginia	14.5	14.8	15.2	2.3	3.0
Wisconsin	74.3	75.3	78.3	1.4	4.0
Wyoming	4.9	5.1	5.4	2.7	6.3
US sum	3,989.1	3,943.2	4,040.9	–1.2	2.5
US published	3,989.1	3,943.2	4,040.9	–1.2	2.5

[a]In addition to private wages, the CBP payroll data cover those government employees who work in government hospitals, federally chartered savings institutions and credit unions, liquor stores, and wholesale liquor establishments.

[b] The BLS data do not cover certain religious elementary and secondary schools because a Supreme Court decision exempts some of these schools from unemployment compensation taxes. The BLS data also exclude college students (and their spouses) who are employed by the school in which they are enrolled and student nurses and interns who are employed by hospitals as part of their training. In half of the states, the BLS data only include nonprofit

TABLE 7-6 Continued

	BLS less Census Bureau			% Difference		
	2001	2002	2003	2001	2002	2003
Alabama	–0.3	0.1	–0.3	–0.7	0.2	–0.5
Alaska	–1.0	–0.8	–0.8	–11.9	–9.7	–9.5
Arizona	2.7	2.0	1.5	4.5	3.3	2.3
Arkansas	–0.3	0.0	–0.5	–1.2	0.1	–1.9
California	–3.2	–2.7	0.5	–0.6	–0.5	0.1
Colorado	0.1	1.1	1.2	0.1	1.6	1.7
Connecticut	–0.6	–1.7	–1.7	–0.9	–2.5	–2.4
Delaware	–1.4	–1.0	–0.9	–9.0	–6.7	–5.8
D.C.	1.3	1.3	1.2	6.1	6.3	5.2
Florida	1.4	2.7	2.1	0.7	1.4	1.1
Georgia	0.3	1.1	0.1	0.3	0.9	0.0
Hawaii	0.6	0.4	0.4	4.9	3.0	2.7
Idaho	0.5	0.4	0.1	3.7	2.9	1.1
Illinois	–5.1	–1.0	–2.8	–2.5	–0.5	–1.4
Indiana	–0.4	0.2	–0.4	–0.5	0.3	–0.4
Iowa	–0.4	–0.2	–0.4	–1.0	–0.7	–1.1

QCEW Private Wages[b]			Yr-Yr % Chg	
2001	2002	2003	2002	2003
7.5	7.5	7.7	0.9	2.3
103.2	102.4	106.3	–0.8	3.9
82.6	81.8	83.5	–1.0	2.0
15.1	15.2	15.4	1.0	1.0
73.0	74.2	76.1	1.7	2.6
5.0	5.2	5.3	3.2	3.8
3,952.2	3,930.8	4,015.8	–0.5	2.2
3,952.2	3,930.8	4,015.8	–0.5	2.2

organizations with four or more employees during 20 weeks in a calendar year. Beginning in 2001, BLS classifies all Native American tribal data under local government; previously, commercial establishments were classified as private.

SOURCES: CBP (County Business Patterns) from the Census Bureau web site: 2003 data released 08/05, downloaded 9/28/05. QCEW (Quarterly Census of Employment and Wages) from flat file downloaded from BLS web site on 9/28/05. Fixed decimals.

State and Local Inc Tax as % of Total Wages			Effective State and Local Tax BLS to Census Bureau Difference (millions of dollars)		
2001	2002	2003	2001	2002	2003
3.70	3.57	3.49	–11	3	–9
0.00	0.00	0.00	0	0	0
2.80	2.57	2.56	76	52	38
4.89	4.68	4.53	–15	2	–23
6.67	5.21	5.18	–211	–139	25
3.48	3.96	3.72	2	44	44
5.26	4.48	4.64	–34	–77	–79
5.02	4.47	4.43	–68	–44	–39
2.79	2.27	2.28	36	31	26
0.00	0.00	0.00	0	0	0
4.78	4.46	4.44	16	48	3
5.54	5.26	4.88	34	21	19
6.60	4.93	4.89	31	18	7
3.05	3.17	3.06	–155	–33	–84
4.53	4.33	4.32	–17	11	–15
4.53	4.24	4.15	–16	–11	–17

Continued

TABLE 7-6 Continued

	BLS less Census Bureau			% Difference		
	2001	2002	2003	2001	2002	2003
Kansas	–0.3	–0.1	–0.7	–0.9	–0.3	–2.0
Kentucky	0.7	1.3	1.3	1.8	3.0	2.9
Louisiana	–0.9	–0.9	–1.4	–2.1	–2.0	–3.0
Maine	–0.1	0.0	0.1	–0.8	0.2	0.6
Maryland	–0.8	–0.1	–1.2	–1.1	–0.2	–1.5
Massachusetts	–4.3	–1.5	0.5	–3.2	–1.2	0.4
Michigan	1.0	0.5	1.3	0.7	0.4	0.9
Minnesota	–2.4	–1.5	–1.9	–2.9	–1.7	–2.2
Mississippi	–0.1	0.2	–0.2	–0.4	1.0	–0.8
Missouri	–1.3	–0.4	–0.9	–1.7	–0.6	–1.2
Montana	0.2	0.3	0.3	2.4	3.7	4.3
Nebraska	–0.2	–0.7	–1.4	–1.0	–3.3	–6.1
Nevada	2.1	1.0	1.2	7.7	3.4	4.0
New Hampshire	0.6	0.4	0.8	3.2	2.3	4.1
New Jersey	–8.3	–5.4	–4.5	–5.4	–3.5	–2.9
New Mexico	0.7	0.8	0.7	4.9	5.5	4.2
New York	–8.9	–7.7	–6.7	–2.6	–2.3	–2.0
North Carolina	–1.1	0.0	–1.5	–1.1	0.0	–1.4
North Dakota	–0.1	0.0	–0.1	–1.4	–0.4	–0.9
Ohio	–2.8	–0.6	–0.9	–1.8	–0.4	–0.6
Oklahoma	–1.1	–1.3	–0.8	–3.2	–3.9	–2.3
Oregon	–0.1	0.1	0.1	–0.2	0.3	0.3
Pennsylvania	–2.3	–0.2	–1.1	–1.4	–0.1	–0.6
Rhode Island	–0.1	–0.1	–0.3	–1.0	–0.9	–2.2
South Carolina	–1.3	–1.1	–0.6	–3.0	–2.5	–1.4
South Dakota	–0.1	–0.1	–0.1	–1.9	–1.4	–0.6
Tennessee	–0.2	0.5	0.6	–0.3	0.7	0.9
Texas	4.3	3.9	2.5	1.5	1.4	0.9
Utah	–0.2	–0.4	–0.6	–0.7	–1.6	–2.4
Vermont	0.2	0.1	0.1	2.6	1.5	0.8
Virginia	0.7	0.7	0.3	0.7	0.7	0.3
Washington	–3.9	–1.3	–7.1	–4.5	–1.6	–7.9
West Virginia	0.6	0.4	0.1	4.1	2.8	0.9
Wisconsin	–1.3	–1.1	–2.2	–1.8	–1.5	–2.8
Wyoming	0.1	0.1	0.0	1.5	2.0	–0.5
US sum	–36.9	–12.4	–25.1	–0.9	–0.3	–0.6
US published	–36.9	–12.4	–25.1	–0.9	–0.3	–0.6

SOURCES: CBP (County Business Patterns) from Census Bureau web site: 2003 data released 08/05, downloaded 9/28/05. QCEW (Quarterly Census of Employment and Wages) from flat file downloaded from BLS web site on 9/28/05. Fixed decimals.

State and Local Inc Tax as % of Total Wages			Effective State and Local Tax BLS to Census Bureau Difference (millions of dollars)		
2001	2002	2003	2001	2002	2003
4.72	4.32	4.19	–14	–4	–29
6.24	6.20	6.21	47	79	80
3.19	3.16	3.12	–30	–29	–44
6.54	5.91	5.75	–7	2	5
7.81	7.42	7.35	–65	–11	–88
6.29	5.12	5.46	–269	–76	25
4.14	3.75	3.57	42	20	47
5.05	5.44	5.25	–123	–79	–101
3.29	3.18	3.16	–3	8	–6
4.66	4.25	4.17	–59	–19	–37
5.32	4.79	4.83	9	13	16
4.57	4.16	3.97	–10	–30	–56
0.00	0.00	0.00	0	0	0
0.35	0.30	0.22	2	1	2
4.44	3.75	3.67	–367	–202	–166
3.94	3.99	3.80	29	33	26
7.93	6.96	6.96	–709	–536	–468
5.89	5.59	5.42	–67	–3	–81
2.47	2.20	2.12	–2	–1	–1
6.28	6.30	5.97	–175	–37	–54
5.22	5.00	4.98	–55	–66	–38
7.31	7.23	7.47	–7	9	9
4.90	4.58	4.54	–115	–10	–49
5.46	4.88	4.65	–7	–6	–15
4.35	4.14	4.09	–57	–46	–25
0.00	0.00	0.00	0	0	0
0.23	0.16	0.13	0	1	1
0.00	0.00	0.00	0	0	0
5.14	4.64	4.60	–9	–20	–29
4.66	4.78	4.26	9	5	3
5.61	4.65	4.91	38	32	13
0.00	0.00	0.00	0	0	0
5.09	5.14	5.03	30	22	7
5.80	5.39	5.44	–77	–60	–119
0.00	0.00	0.00	0	0	0
—	—	—	–2,353	–1,084	–1,276
—	—	—	—	—	—

public policy and business administration. Specifically, regional data affect federal government fund allocations to states, analysis of state government tax efforts, state and local government tax and revenue planning, business analyses of the size of markets, and the extent of the safety net at the state level. For example, the Medicaid program uses BEA per capita personal income in the Federal Medicaid Assistance Percentage formula to determine the federal share of payments for each state. Since wages and salaries and wage-related components account for about two-thirds of personal income, the level of differences among the states—varying from $2.5 billion in Texas to –$7.1 billion in Washington for 2003—would have a significant impact on the federal share of Medicaid payments for each state.

In New York, the $1.2 billion dollar difference in growth in wages and salaries from 2001 to 2002 between BLS and the Census Bureau series, shown in Table 7-6, would amount to about a $173 million difference in projected state and local government income taxes received. New Jersey would have a $165 million projected difference in income taxes, and Massachusetts would have a $193 million projected difference.

If the difference between BLS and the Census Bureau reflects coverage differences, then, for example, the 10.4 percent average difference in Alaska for the years 2001 through 2003 would reflect the percentage of the workforce that is not covered by unemployment insurance. This information would be very useful to those officials interested in the extent of the unemployment insurance safety net among the states.

Finally, the differences may also reflect changes in coverage between the programs. For example, beginning in 2004, the Washington State unemployment insurance program no longer covers exercised stock options in its definition of covered wages and salaries, whereas the Census Bureau wages continue to cover exercised stock options for all states. Thus an increase in the differences between the two programs is expected beginning in 2004. Access to microdata in the programs would allow identification of the amount of the exercised stock options in the state.

Regional and Industry Influences Combined: County Business Patterns by Industry Compared with BLS QCEW Data

BEA uses BLS QCEW data in its initial estimates of wages, since it is the most comprehensive and earliest available data. This choice is made because the CBP data become available later. These data, however, are used as a check on the initial estimates as well as a data source for certain areas that are not covered by QCEW. Although the two data series have a large amount of overlap, at the NAICS sector (two-digit) level of wage detail, CBP data and QCEW data can show different trends for certain industries. In Table 7-7, the mean absolute difference in average annual growth between 1998 and 2002 across industries is 0.9 percentage points. Table 7-8 shows larger variations across industries with respect to one-year growth rates: mean absolute differences in growth rates are 1.0, 2.1, 2.0, and 1.2 percentage points for 1999 through 2002, respectively.[9]

For individual industries, there is a considerable difference in growth rates as well. Between 1998 and 2002 the absolute difference in growth rates can vary as much as 2.2 percentage points, as is the case for the information sector (51), and as little as 0.2 percentage points, as in the health care and social assistance sector (62). One-year growth rates show larger differences. In 2002 there was a 5.6 percentage point difference in growth for the administrative, support, waste management, and remediation services sector (56), and in 2001 there was a 6.5 percentage point difference for the arts, entertainment, and recreation sector (71). In 2000, there was a 9.0 percentage point difference in the growth rate for the information sector (51) and a 6.7 percentage point difference for the forestry, fishing, hunting, and agriculture support sector (11). In the latter case, CBP data show a decline of 2.7 percent, while QCEW data show an increase of 4.0 percent.

Employment Differences and BEA International Accounts

BEA collects data on the activity of multinational firms at the enterprise level. In the early 1990s a study was conducted to compile establishment-based data for foreign-owned establishments in the United States, the results of which are included in *Foreign Direct Investment in the United States: Establishment Data for 1987*. For this study, BEA shared its confidential enterprise-level data with BLS and the Census Bureau so that each could determine the relevant set of establishments. There was no interaction between the Census Bureau and BLS in the compilation of each list. As shown in Table 7-9, BLS identified 3 percent more establishments for all industries than the Census Bureau, and the corresponding BLS employment level is 4 percent less than the corresponding Census

[9]All of the QCEW data in Table 7-7 are based on the private sector in order to better match the survey population of the CBP. In addition, it should be noted that the 2001 growth rates are based on 2000 levels that are backcasted for QCEW; that is, all QCEW data prior to 2001 are backcasted for NAICS. For a description of the backcasting procedure, see Morisi (2003). Many of the larger differences between the data series during the period of backcasting are in services sectors.

TABLE 7-7 Private Annual Payroll Data—Census Bureau (CBP) and BLS (QCEW): Levels

NAICS Code	Industry Code Description	1998 CBP	QCEW	$ Diff.	% Diff.
	Total	3,309.4	3,338.7	29.3	0.9
11	Forestry, fishing, hunting, and agric. support	4.7	21.5	16.8	362.0
21	Mining	21.9	29.5	7.5	34.3
22	Utilities	38.1	34.5	–3.6	–9.4
23	Construction	198.5	202.9	4.4	2.2
31	Manufacturing	607.3	678.9	71.6	11.8
42	Wholesale trade	233.9	238.5	4.6	2.0
44	Retail trade	260.3	294.3	34.0	13.1
48	Transportation and warehousing	108.6	128.7	20.2	18.6
51	Information	146.8	160.1	13.2	9.0
52	Finance and insurance	290.0	279.2	–10.8	–3.7
53	Real estate, rental, and leasing	49.9	55.8	5.9	11.9
54	Professional, scientific, and technical services	277.6	309.1	31.4	11.3
55	Management of companies and enterprises	175.6	105.4	–70.3	–40.0
56	Admin, support, waste management, remediation services	163.7	152.2	–11.5	–7.0
61	Educational services	52.3	47.7	–4.6	–8.8
62	Health care and social assistance	395.5	358.3	–37.2	–9.4
71	Arts, entertainment, and recreation	36.0	37.9	1.8	5.1
72	Accommodation and food services	109.6	117.6	8.0	7.3
81	Other services (except public administration)	96.0	82.3	–13.7	–14.2
	Mean (excluding Total and 11)				1.9

Bureau level. However, at the individual industry levels, the differences can be substantial. The average percentage difference in the number of establishments (fifth column in the table), without regard to direction of the difference, is 24 percent and the corresponding average difference in employment is 7.7 percent. Inasmuch as the table reflects data classified under the Standard Industrial Classification (SIC) system, the movement to NAICS is likely to have affected these estimates. For example, under the SIC system, auxiliaries in manufacturing (mainly head offices) were included in the manufacturing sector as a separate industry, and while the Census Bureau followed that classification, BLS dispersed the auxiliaries into the other manufacturing industries. Under NAICS, auxiliaries are still separately identified, but they are now placed in a sector entitled

1999			
CBP	QCEW	$ Diff.	% Diff.
3,554.7	3,594.7	40.1	1.1
4.8	22.3	17.5	363.6
21.0	28.1	7.1	33.8
39.4	35.5	–3.9	–9.8
219.1	224.2	5.1	2.3
625.5	702.4	76.9	12.3
250.0	255.6	5.6	2.2
281.9	314.3	32.4	11.5
116.7	137.7	21.0	18.0
170.3	186.2	15.9	9.4
313.2	303.0	–10.3	–3.3
54.1	59.4	5.3	9.9
311.2	362.5	51.3	16.5
192.4	113.3	–79.1	–41.1
183.1	167.1	–16.0	–8.7
56.9	51.2	–5.7	–10.0
409.2	372.1	–37.1	–9.1
39.4	41.0	1.6	4.0
116.9	125.7	8.8	7.5
102.0	86.9	–15.1	–14.8
			1.7

Continued

"Management of Companies and Enterprises"; thus, they are no longer in the sector of the establishments that they serve. Indeed, one of the major differences in the number of manufacturing establishments that can be found in the 1997 economic census between the SIC system and NAICS is due to the different treatment of auxiliary establishments.

A more recent study examined differences in establishments and employment in the area of research and development (R&D) performed by U.S. firms. The NSF through the Census Bureau collects data on the R&D expenditures of U.S. firms. BEA collects data on the R&D expenditures of U.S. and foreign multinational companies. The BEA/Census Bureau/NSF R&D link project was a study to determine whether an integrated data set on U.S. R&D performance and funding could be created by linking the Census Bureau data on the R&D activity of all U.S. companies with BEA

TABLE 7-7 Continued

NAICS		2000			
Code	Industry Code Description	CBP	QCEW	$ dif.	% dif.
	Total	3,879.4	3,889.0	9.6	0.2
11	Forestry, fishing, hunting, and agric. support	4.7	23.2	18.5	395.7
21	Mining	22.1	29.7	7.6	34.3
22	Utilities	40.7	37.9	–2.7	–6.7
23	Construction	239.9	245.8	5.9	2.4
31	Manufacturing	644.0	743.8	99.8	15.5
42	Wholesale trade	270.1	276.8	6.7	2.5
44	Retail trade	302.6	335.8	33.2	11.0
48	Transportation and warehousing	125.6	147.2	21.6	17.2
51	Information	209.4	212.3	2.9	1.4
52	Finance and insurance	346.8	333.8	–13.0	–3.7
53	Real estate, rental, and leasing	59.2	64.4	5.2	8.8
54	Professional, scientific, andtechnical services	362.0	395.4	33.4	9.2
55	Management of companies and enterprises	211.4	124.0	–87.4	–41.4
56	Admin, support, waste management, remediation services	210.3	185.4	–24.9	–11.8
61	Educational services	61.9	55.6	–6.4	–10.3
62	Health care and social assistance	431.4	394.7	–36.8	–8.5
71	Arts, entertainment and recreation	43.2	45.1	1.9	4.4
72	Accommodation and food services	125.6	134.2	8.6	6.9
81	Other services (except public administration)	109.9	92.7	–17.2	–15.6
	Mean (excluding Total and 11)				0.9

NOTE: BLS QCEW data prior to 2001 have been backcasted to NAICS 2002 using NAICS reports from employers in the first quarter of 2001. Data for 2001 and 2002 also use NAICS 2002. CBP data are based on NAICS 1997.

data. This project was conducted under the authority of the International Investment and Trade Act and CIPSEA. No Title 26 data were used in the linking operation or subsequent tabulations or reports for this study; neither BEA nor the Census Bureau data sets used for the project contained such data, as all original FTI were replaced by respondent data for the Census Bureau surveys being linked. The Census Bureau informed the IRS of the project to alleviate any questions or concerns the IRS might have.

The project demonstrated that it is feasible to link the Census Bureau and BEA survey data, and that by linking the data an integrated data set on the domestic and international dimensions of R&D can be created. Table 7-10 compares the NSF data with the BEA data for U.S. parent companies and therefore examines only a subset of U.S. firms—U.S. affiliates of foreign companies were not included. Despite the smaller BEA uni-

2001				2002			
CBP	QCEW	$ dif.	% dif.	CBP	QCEW	$ dif.	% dif.
3,989.1	3,952.2	–36.9	–0.9	3,943.2	3,930.8	–12.4	–0.3
4.8	23.6	18.8	393.2	5.0	24.1	19.2	385.0
25.0	31.9	6.9	27.7	24.0	30.6	6.6	27.5
41.9	39.3	–2.6	–6.2	41.8	39.9	–1.9	–4.7
247.2	260.2	13.0	5.2	247.3	260.8	13.5	5.5
617.7	704.1	86.4	14.0	580.4	670.7	90.3	15.6
275.9	279.6	3.7	1.3	262.5	276.6	14.1	5.4
314.8	344.1	29.3	9.3	320.7	348.9	28.2	8.8
129.5	149.8	20.3	15.6	127.3	146.9	19.6	15.4
207.1	205.8	–1.3	–0.7	188.1	188.8	0.7	0.4
373.6	359.4	–14.2	–3.8	372.7	356.4	–16.3	–4.4
64.0	66.9	2.9	4.6	65.2	68.8	3.6	5.5
374.4	403.7	29.3	7.8	368.8	390.5	21.7	5.9
213.1	118.4	–94.8	–44.5	204.8	117.5	–87.3	–42.6
221.4	189.1	–32.3	–14.6	212.2	191.8	–20.4	–9.6
67.1	60.4	–6.7	–10.0	72.0	64.7	–7.3	–10.1
465.7	425.5	–40.3	–8.6	499.2	456.0	–43.1	–8.6
46.1	45.2	–0.9	–2.0	47.7	47.1	–0.7	–1.4
128.6	138.1	9.5	7.4	131.1	142.2	11.1	8.5
115.2	97.7	–17.5	–15.2	118.9	101.0	–17.9	–15.1
			–0.7				0.1

SOURCES: Census Web site (5/23/05): http://censtats. census.gov/cgi-bin/cbpnaic/cbpsel.pl; and BLS Web site (5/24/05): ftp://ftp.bls.gov/pub/special.requests/cew/.

verse, the table shows that in some industries BEA data indicate a far higher level of R&D expenditures than that for all U.S. firms—pharmaceuticals and medicines, for example. At the micro level, there were 11 cases in which the BEA and Census data for total R&D spending for the matched U.S. parent companies differed by more than $500 million. There is a substantial difference in the collected data for manufacturing and nonmanufacturing. The substantially lower number for nonmanufacturing may result from the fact that R&D expenditures for nonmanufacturing firms are relatively more difficult to define and identify; so this area is more likely to be affected by differences in treatment.

The study also demonstrated some of the main benefits of data sharing—in the improvement of sample frames and the quality of reported data. For example, as a result of the project, the Census Bureau added over 500 companies to the sample for the Survey of Industrial Research

TABLE 7-8 Private Annual Payroll Data—Census Bureau (CBP) and BLS (QCEW): Growth

		Growth Rate (%)			
		1999			
NAICS Code	Industry Code Description	CBP	QCEW	Diff.	Ab. Diff.
	Total	7.4	7.7	0.3	0.3
11	Forestry, fishing, hunting, and agric. support	3.4	3.8	0.4	0.4
21	Mining	–4.4	–4.7	–0.3	0.3
22	Utilities	3.3	2.8	–0.5	0.5
23	Construction	10.4	10.5	0.1	0.1
31	Manufacturing	3.0	3.5	0.5	0.5
42	Wholesale trade	6.9	7.2	0.3	0.3
44	Retail trade	8.3	6.8	–1.5	1.5
48	Transportation and warehousing	7.5	6.9	–0.5	0.5
51	Information	16.0	16.3	0.4	0.4
52	Finance and insurance	8.0	8.5	0.5	0.5
53	Real estate, rental, and leasing	8.4	6.5	–2.0	2.0
54	Professional, scientific, and technical services	12.1	17.3	5.2	5.2
55	Management of companies and enterprises	9.5	7.5	–2.0	2.0
56	Admin, support, waste management, remediation services	11.9	9.8	–2.0	2.0
61	Educational services	8.8	7.4	–1.4	1.4
62	Health care and social assistance	3.5	3.9	0.4	0.4
71	Arts, entertainment, and recreation	9.4	8.2	–1.1	1.1
72	Accommodation and food services	6.7	6.9	0.2	0.2
81	Other services (except public administration)	6.2	5.5	–0.7	0.7
	Minimum			–2.0	0.1
	Mean			–0.2	1.0
	Maximum			5.2	5.2

and Development. For more information, see a report on the findings of the project—"Research and Development Link Project: Final Report" at www.bea.gov/bea/di/FinalReportpublic.pdf.

How Data Sharing Could Help

A large part of BEA's job is adjusting the various data for differences in timing, concepts, and definitions. However, this is often difficult because, for the most part, BEA does not have access to the underlying

2000			
CBP	QCEW	Diff.	Ab. Diff.
9.1	8.2	–0.9	0.9
–2.7	4.0	6.7	6.7
5.3	5.7	0.4	0.4
3.3	6.9	3.6	3.6
9.5	9.6	0.1	0.1
2.9	5.9	2.9	2.9
8.0	8.3	0.2	0.2
7.3	6.8	–0.5	0.5
7.6	6.9	–0.7	0.7
23.0	14.0	–9.0	9.0
10.7	10.2	–0.5	0.5
9.5	8.4	–1.1	1.1
16.3	9.1	–7.2	7.2
9.9	9.4	–0.4	0.4
14.8	11.0	–3.9	3.9
8.9	8.5	–0.3	0.3
5.4	6.1	0.6	0.6
9.6	10.1	0.4	0.4
7.4	6.8	–0.6	0.6
7.8	6.7	–1.0	1.0
		–9.0	0.1
		–0.6	2.1
		6.7	9.0

Continued

microdata. If armed with full data-sharing capability, BEA, BLS, and the Census Bureau could explore and resolve differences in the activities of major companies or in their classification to various industries and regions. The agencies could also compare data to investigate and resolve persistent differences, such as the reporting of bonuses and stock options, the capitalization of computer investment, the impact of differences in timing, and the differences in company practices with respect to the writing-down of inventories or to the treatment of pensions and other fringe benefits.

TABLE 7-8 Continued

		Growth Rate (%)			
		2001			
NAICS Code	Industry Code Description	CBP	QCEW	Diff.	Ab. Diff.
	Total	2.8	1.6	–1.2	1.2
11	Forestry, fishing, hunting, and agric. support	2.3	1.8	–0.5	0.5
21	Mining	13.2	7.7	–5.6	5.6
22	Utilities	3.2	3.7	0.5	0.5
23	Construction	3.0	5.9	2.8	2.8
31	Manufacturing	–4.1	–5.3	–1.3	1.3
42	Wholesale trade	2.1	1.0	–1.1	1.1
44	Retail trade	4.0	2.5	–1.6	1.6
48	Transportation and warehousing	3.1	1.8	–1.3	1.3
-51	Information	1.1	–3.1	–2.0	2.0
52	Finance and insurance	7.7	7.7	–0.1	0.1
53	Real estate, rental, and leasing	8.1	3.9	–4.2	4.2
54	Professional, scientific, and technical services	3.4	2.1	–1.3	1.3
55	Management of companies and enterprises	0.8	–4.5	–5.4	5.4
56	Admin, support, waste management, remediation services	5.3	2.0	–3.3	3.3
61	Educational services	8.4	8.7	0.4	0.4
62	Health care and social assistance	7.9	7.8	–0.1	0.1
71	Arts, entertainment, and recreation	6.8	0.3	–6.5	6.5
72	Accommodation and food services	2.4	2.9	0.5	0.5
81	Other services (except public administration)	4.9	5.3	0.5	0.5
	Minimum			–6.5	0.1
	Mean			–1.5	2.0
	Maximum			2.8	6.5

NOTE: BLS QCEW data prior to 2001 have been backcasted to NAICS 2002 using NAICS reports from employers in the first quarter of 2001. Data for 2001 and 2002 also use NAICS 2002. CBP data are based on NAICS 1997.

The limited access also affects BEA's ability to study observed anomalies in the Census Bureau data. The following are some examples of observations that BEA would like to study.

• There are substantial differences in the reported payrolls from the Census Bureau and BLS, by area. For example, in Washington state, between 1999 and 2000, the Census Bureau reports an 11.5 percent increase (more than $9 billion) while BLS reports a 6.2 percent increase ($5 billion). There are many possible reasons for the discrepancy, and data sharing

2002				1998-2002			
CBP	QCEW	Diff.	Ab. Diff.	CBP	QCEW	Diff.	Ab. Diff.
–1.2	–0.5	0.6	0.6	4.5	4.2	–0.3	0.3
3.9	2.2	–1.7	1.7	1.7	2.9	1.2	1.2
–4.2	–4.3	–0.1	0.1	2.2	0.9	–1.3	1.3
–0.2	1.4	1.7	1.7	2.4	3.7	1.3	1.3
0.0	0.3	0.2	0.2	5.6	6.5	0.8	0.8
–6.0	–4.7	1.3	1.3	–1.1	–0.3	0.8	0.8
–4.8	–1.1	3.8	3.8	2.9	3.8	0.8	0.8
1.9	1.4	–0.5	0.5	5.4	4.3	–1.0	1.0
–1.7	–1.9	–0.2	0.2	4.0	3.4	–0.7	0.7
–9.2	–8.3	0.9	0.9	6.4	4.2	–2.2	2.2
–0.2	–0.8	–0.6	0.6	6.5	6.3	–0.2	0.2
1.9	2.8	0.9	0.9	6.9	5.4	–1.6	1.6
–1.5	–3.3	–1.8	1.8	7.4	6.0	–1.3	1.3
–3.9	–0.8	3.2	3.2	3.9	2.7	–1.2	1.2
–4.2	1.4	5.6	5.6	6.7	6.0	–0.7	0.7
7.2	7.1	–0.1	0.1	8.3	7.9	–0.4	0.4
7.2	7.2	0.0	0.0	6.0	6.2	0.2	0.2
3.4	4.0	0.6	0.6	7.3	5.6	–1.7	1.7
1.9	3.0	1.1	1.1	4.6	4.9	0.3	0.3
3.2	3.4	0.2	0.2	5.5	5.2	–0.3	0.3
		–1.8	0.0			–2.2	0.2
		0.7	1.2			–0.4	0.9
		5.6	5.6			1.3	2.2

SOURCES: Census Bureau web site (5/23/05): http://censtats. census.gov/cgi-bin/cbpnaic/cbpsel.pl and BLS Web site (5/24/05): ftp://ftp.bls.gov/pub/special.requests/cew/.

with access to tax data would help get at the cause. For example, one source of the difference could be differences in the recording of stock options. By knowing the companies in the state, it would be possible to check with firm reports about reported stock options and thereby reconcile any difference. Such huge differences in the payroll numbers affect the estimation of GDI.

• BEA obtains monthly data from the Census Bureau for the manufacturing sector based on the M3 (Manufacturers' Shipments, Inventories and Orders) survey. However, because participation in this survey is vol-

TABLE 7-9 Comparison of the Census Bureau and BLS Data for Foreign-Owned Establishments, 1992

	Census Bureau		BLS		BLS Value as Percentage of Census Bureau Value	
	Number of Establishments	Number of Employees	Number of Reporting Units	Number of Employees	Number of Establishments	Number of Employees
All industries	102,958	4,944,157	106,041	4,747,637	103	96
Agricultural services, forestry, and fishing[a]	139	5,814	139	4,265	100	73
Mining		120,782	1,640	102,814	102	85
Construction	3,322	2,305	90,866	189	97	
Manufacturing	81	2,004,947	13,076	1,930,135	102	96
Transportation and public utilities[b]	231,638	3,792	222,999	97	96	
Wholesale trade	91	513,012	34,999	491,578	186	96
Retail trade	3,190	26,756	853,158	71	100	
Finance, insurance, and real estate	401,018	9,558	360,287	83	90	
Services	2,775	12,899	676,091	85	94	
Other[c]	9	1,116	569	3,137	n.m.	n.m.

NOTE: n.m. = not meaningful.

[a]Excludes agricultural production of crops and livestock.

[b]The Census Bureau data exclude railroad transportation.

[c]For the Census Bureau: consists of private education and noncommerical establishments; for BLS: consists of nonclassifiable establishments.

SOURCES: The Census Bureau data: Foreign Direct Investment in the United States: Establishment Data for 1992 available on BEA's Web site at http://www.bea.gov/bea/ai1.htm#BEACENS. BLS data: BLS news release: "Employment and Wages in Foreign-Owned Businesses in the United States, Fourth Quarter 1992," October 1996.

TABLE 7-10 Comparison of NSF R&D Expenditures by All U.S. Companies with BEA R&D Expenditures by U.S. Parent Companies, 2001 (millions of dollars or percentage)

	NSF: All U.S. Companies	BEA: U.S. Parent Companies	Parent Companies as Percentage of all U.S. Companies
All industries	198,505	143,017	72
Manufacturing	120,705	115,118	95
Food	1,819	914	50
Beverage and tobacco products	152	469	309
Textiles, apparel, and leather	(D)	125	n.a.
Wood products	182	(D)	n.a.
Paper, printing and support activities	(D)	(D)	n.a.
Petroleum and coal products	(D)	1,002	n.a.
Chemicals	17,892	31,927	178
Basic chemicals	1,876	1,742	93
Resin, synthetic rubber, fibers, and filament	(D)	2,972	n.a.
Pharmaceuticals and medicines	10,137	23,169	229
Other chemicals	(D)	4,045	n.a.
Plastics and rubber products	(D)	929	n.a.
Nonmetallic mineral products	990	339	34
Primary metals	485	484	100
Fabricated metal products	1,599	554	35
Machinery	6,404	8,561	134
Computer and electronic products	47,079	38,356	81
Computers and peripheral equipment	(D)	7,727	n.a.
Communications equipment	15,507	14,526	94
Semiconductor and other electronic components	14,358	11,114	77
Navigational, measuring, electromedical, and control instruments	12,947	4,158	32
Other computer and electronic products	(D)	832	n.a.
Electrical equipment, appliances, and components	4,980	2,008	40
Transportation equipment	25,965	25,147	97
Motor vehicles, trailers, and parts	(D)	18,183	n.a.
Other	(D)	6,964	n.a.
Furniture and related products	301	128	43
Miscellaneous manufacturing	6,606	2,570	39
Nonmanufacturing	77,799	27,899	36
Mining, extraction, and support activities	(D)	411	n.a.
Utilities	133	59	44

Continued

TABLE 7-10 Continued

	NSF: All U.S. Companies	BEA: U.S. Parent Companies	Parent Companies as Percentage of all U.S. Companies
Construction	320	(D)	n.a.
Trade	24,372	(D)	n.a.
Transportation and warehousing	1,848	12	1
Information	(D)	9,514	n.a.
Publishing	13,760	6,452	47
Newspaper, periodical, book, and database	649	(D)	n.a.
Software	13,111	(D)	n.a.
Broadcasting and telecommunications	(D)	796	n.a.
Telecommunications	(D)	782	n.a.
Other	(D)	14	n.a.
Other information	(D)	2,266	n.a.
Finance, insurance, and real estate	(D)	624	n.a.
Professional, scientific, and technical services	27,704	10,348	37
Architectural, engineering, and related services	3,386	18	1
Computer systems design and related services	9,154	8,929	98
Other	15,164	1,401	9

NOTES: (D) = suppressed to avoid disclosure of data of individual companies; n.a. = not available.

SOURCES: R&D spending by all U.S. companies: *Research and Development in Industry: 2001* available on the NSF web site at http://www.nsf.gov/statistics/nsf05305/htmstart.htm; R&D spending by U.S. parent companies: U.S. Direct Investment Abroad: Operations of U.S. Parent Companies and Their Foreign Affiliates, Revised 2001 Estimates available on the BEA's web site at http://www.bea.gov/bea/ai/iidguide.htm#link12b.

untary and some firms decide not to participate, BEA does not know the extent of participation. A recent example is the decision by a major producer of semiconductors to terminate its participation, which represented a huge erosion in the representativeness of the surveys. The Annual Survey of Manufacturers, however, is mandatory. Thus BEA must wait until the annual data are available before it can check the estimates based on the monthly data. If BEA had access to the M3 data, then it could identify the firms responsible for missing data and possibly estimate the missing information from publicly available sources such as company reports. Publicly available sales data from company reports could aid in the estimation of missing shipment data from a company that did not provide such information in the M3 survey. The ability to estimate such missing information would serve to reduce revisions to GDP.

- Relatedly, the M3 surveys provide inventories data to BEA, and BEA does not always know whether a company has reported an inventory adjustment to the Census Bureau in the same way that the company has entered it on its financial accounts. A few years ago the press reported a major write-down of inventory by a major producer of information technology equipment. Without knowing how the firm entered the write-down in its Census Bureau report, BEA had no way of checking if the inventory adjustment was accurately reflected both on the product side of the national accounts (inventories is a component of investment) and on the income side (the valuation of inventories affects corporate profits). In fact, there was a large adjustment to inventories.

How Data Sharing Would Help Cope with Disasters

The massive destruction wrought by Hurricane Katrina is having a significant impact on the ability of the statistical agencies to collect economic data in the affected regions. As a result of the disappearance of sample units, estimates of retail trade, construction, employment and wages, and other components of the principal economic indicators will contain many imputations. Data sharing would allow a combining of data that would enable the statistical agencies to better impute missing values. For example, in the absence of complete business list reconciliation between the Census Bureau and BLS, data sharing would allow one of the agencies to find alternative establishments that might serve as proxies for missing establishments and thereby provide a straightforward imputation. In addition, the Census Bureau sales values might be used by BLS to impute prices for its price indexes. The ability to share data would also enable the statistical agencies to examine each other's establishment-level imputations to see if they suit an agency's needs. For example, the BEA regional program would have access to the BLS establishment imputations for QCEW to see if their needs are met. In short, data sharing would allow the statistical agencies to economize resources to efficiently handle disruptions to the usual production of economic statistics.

REFERENCES

Morisi, T.

2003 Recent changes in the National Current Employment Statistics Survey. *Monthly Labor Review* June.

Moyer, B.C., M.A. Planting, M. Fahim-Nader, and S.K.S. Lum

2004 Preview of the comprehensive revision of the annual industry accounts. *Survey of Current Business* 84(March):38-51.

National Science Foundation, Division of Science Resources Statistics
2005 *Research and Development in Industry: 2001.* (NSF #05-305.) Arlington, VA: National Science Foundation. Available: http://nsf.gov/statistics/nsf05305/front.htm [accessed May 2006].

Strassner, E.H., G.W. Medeiros, and G.M. Smith
2005 Annual industry accounts: Introducing KLEMS input estimates for 1997-2003. *Survey of Current Business* 85(September):34.

U.S. Bureau of Economic Analysis, International Economic Accounts
2005a *Foreign Direct Investment in the United States: Establishment Data for 1992.* (Census data.) Available: http://www.bea.gov/bea/ai1.htm#BEACENS [accessed May 2006].
2005b *U.S. Direct Investment Abroad: Operations of U.S. Parent Companies and Their Foreign Affiliates, Revised 2001.* Available: http://www.bea.gov/bea/ai/iidguide.htm#link12b [accessed May 2006].

U.S. Bureau of Economic Analysis and U.S. Census Bureau
1992 *Foreign Direct Investment in the United States: Establishment Data for 1987.* Washington, DC: U.S. Government Printing Office.

U.S. Census Bureau
2003 *County Business Patterns, NAICS.* Available: http://censtats.census.gov/cgi-bin/cbpnaic/cbpsel.pl [accessed May 2006].

U.S. Census Bureau, Bureau of Economic Analysis, and National Science Foundation Division of Science Resources Statistics
2005 *Statistics Research and Development Data Link Project: Final Report.* Available: http://www.bea.gov/bea/di/FinalReportpublic.pdf [accessed May 2006].

U.S. Department of Labor, Bureau of Labor Statistics
1996 *Employment and Wages in Foreign-Owned Businesses in the United States, Fourth Quarter 1992.* (News release.) Washington, DC: U.S. Department of Labor, Bureau of Labor Statistics.
2002 *Quarterly Census of Employment and Wages Program* Data. Available: ftp://ftp.bls.gov/pub/special.requests/cew/ [accessed May 2006].

8

Using Tax Return Data to Improve Estimates of Corporate Profits

George A. Plesko
University of Connecticut School of Business

This chapter describes the role of tax and financial accounting information in the estimation of corporate profits, summarizes the current dual-reporting requirement that publicly traded firms are subject to, and describes how a broader use of tax return information, coupled with a greater use of financial accounting information, might increase the accuracy of preliminary national income estimates of corporate profits.

MEASURING CORPORATE INCOME

For the purposes of this volume, there are two relevant measures available to assess the profitability of corporations. The first, based on the rules for financial reporting (generally referred to as "book income"), provides a measure of income to the users of financial statement information. The users, under the concepts of financial accounting for publicly traded corporations, are investors, creditors, and any other party needing information to make a decision about whether or not to engage in a business relationship with a firm, but not necessarily able to compel the firm to provide the information.

It is worth noting that foundations of financial accounting and reporting do not explicitly include tax authorities as a user. The reason for this exclusion is that financial accounting disclosures are intended to provide information to those who do not otherwise have the ability to demand information. The separation of audiences, and rules, leads to the second measure of corporate income, based on the Internal Revenue Code. In contrast to the rules of financial reporting, tax reporting removes much of the discretion for the application of rules that are built into financial reporting.

The differences between the amount of income reported to shareholders and the amounts reported to tax authorities, known as book-tax differences, have generated attention both in the press and in policy. However, once a dual measurement system is in place, income differences are a natural occurrence, and the key issue becomes understanding the causes and consequences of the differences. Book-tax differences have been a characteristic of the U.S. reporting system since the inception of the corporate income tax, generating academic interest from the start. For example, Smith and Butters (1949) provide an analysis of book-tax differences for a small sample of corporations active during the 1930s.

SOURCES OF BOOK-TAX DIFFERENCES

Book-tax differences occur because the amount of income reported as earned is based on different concepts and rules under each reporting system. Since the target audience of financial statements is investors and others who need information to make decisions about a company, including whether to invest in the company's equity or debt, companies that issue publicly traded equity or debt securities are required by the Securities and Exchange Commission to file audited financial statements. Such statements must follow generally accepted accounting principles (GAAP), which include an adherence to pronouncements of the Financial Accounting Standards Board (FASB) and other accounting standards.

Tax accounting is designed to administer the U.S. tax laws, with the Internal Revenue Service (IRS) the primary audience for tax filings. In contrast to GAAP, tax rules can change frequently, depending on legislative initiatives, and are not necessarily designed to present as consistent a definition of income over time as are financial accounting rules.

An important element of financial accounting is the amount of discretion left to the corporation in implementing GAAP in their business. For example, in determining the useful life and depreciation pattern of a capital asset, depreciation schedules of the same asset can vary by company and by usage and usually follow a straight-line pattern. Tax depreciation is determined by the Internal Revenue Code and leaves less discretion to the company. The lack of discretion in the tax code is intended to lead to more uniform application of the tax system.

Differences between tax and financial measures of income can arise from two types of measurement differences in the accounting systems: temporary and permanent. Temporary (timing) differences occur when both tax and financial reporting recognize the same total amount of income or expense, but they do so either over different time periods or in different patterns over the same period. Timing differences arise not only from the different reporting rules under each system, but also because GAAP allows managers greater discretion in determining the amounts of income and expense in each period than does the tax system.

Permanent differences in the measures of income arise when a particular item of income or expense is recognized under one system but not the other. For example, tax-exempt interest on municipal bonds is included in book income but not in the determination of tax net income.

Both temporary and permanent differences are reported in corporations' financial statements. Under FASB's Statement of Financial Accounting Standards Number 109, corporations report a total amount of tax liability based on current-year financial reporting income, delineating the portion currently owed to the government from that which is deferred due to differences in income and expense recognition between the two methods. If the deferred portion is positive, a deferred tax liability is created, representing the amount of taxes not paid on financial statement income during this period because of temporary differences reducing taxable income below book income. Such is usually the case in the short term with depreciation, as more deductions are taken for tax purposes during the early years of an asset's life than are recognized as expenses for book purposes. The deferred tax liability associated with the asset on a corporation's financial statements represent the (undiscounted) amount of tax to be paid in the future relative to future book earnings when the tax depreciation deductions fall below the book depreciation expense. In contrast to deferred tax liabilities, deferred tax assets are created when more taxes are paid than would be paid if financial reporting income were used to base tax liability, and they represent a financial claim on the government for taxes paid ahead of time relative to financial reporting.

Permanent differences, such as the effect of tax-exempt interest, never reverse and therefore do not create deferred tax assets or liabilities. Corporations account for permanent differences in a separate financial disclosure in the tax footnote of their financial statements.

Book-tax differences have generated increased attention since the 1999 Treasury Department report on tax shelters and related testimony (Talisman, 2000), in which reporting differences were suggested as evidence of increased tax sheltering by corporations. The Joint Committee on Taxation (2003) report on Enron provided additional evidence of tax sheltering behavior and of reporting differences for book and tax purposes.

BEA METHODOLOGY

The methodology used by the Bureau of Economic Analysis (BEA) explicitly incorporates a combination of publicly available information drawn from both financial statements and tax return data to estimate corporate income. The concept of corporate income in the national accounts is closer to tax than book income, but the time lag in the availability of tax data, described below, requires BEA to forecast current income based on past tax data and recent financial reports. The recent attention to the role of book-tax differences in affecting the estimate of corporate profits is related to the occurrence of large, unexpected differences in the growth rate of each income measure in the 1990s and early 2000s. The sharp revisions in estimated corporate profits have been discussed by others, including the Congressional Budget Office (2005), Mead et al. (2004), and Patrick (2001).

The rules governing financial reporting by public corporations, specifically Financial Accounting Standard 109 called "Accounting for Income Taxes" (Financial Accounting Standard Board, 1992), require firms to reconcile their book measure of tax liability to their actual liability. Furthermore, income tax reporting requires firms to reconcile their taxable income to their book income on Form 1120. As a result, the underlying economic activity of a firm is reported using two distinct measurement systems and, in theory, allows for a better understanding of the financial position of a firm.

Ideally, the information in the tax and financial statements is complementary, since each provides a unique measure of the same economic activities. Furthermore, research suggests that information about book-tax differences, as reported in firms' tax footnote, is useful to investors (Hanlon et al., 2005) in predicting their future performance.

However, a problem occurs when the components of the tax disclosure are examined at the firm level and a user tries to determine specific information, such as taxable income and a company's tax payments. Financial disclosures have been found to be unsatisfactory in providing inferences about the tax attributes of a firm (see Hanlon, 2003; Plesko, 2000a, 2003).

Figure 8-1 is a schematic of these reporting relations. Until 2004, the reconciliation of income for tax reporting purposes was reported on the Schedule M-1 of the corporate income tax return Form 1120. An analysis of the M-1's shortcomings outlined by Mills and Plesko (2003) shows that these reconciliations were of little practical value to the IRS. As a result, for tax year 2004 and after, the Schedule M-1 has been replaced by Schedule M-3 for larger corporations. The efficacy of the Schedule M-3 has yet to be assessed.

The importance of these issues, from an empirical view, is that the financial statement information that BEA relies on will misrepresent the taxable income numbers it is trying to infer if book-tax differences are large and changing in ways not observable in the financial data. Recent

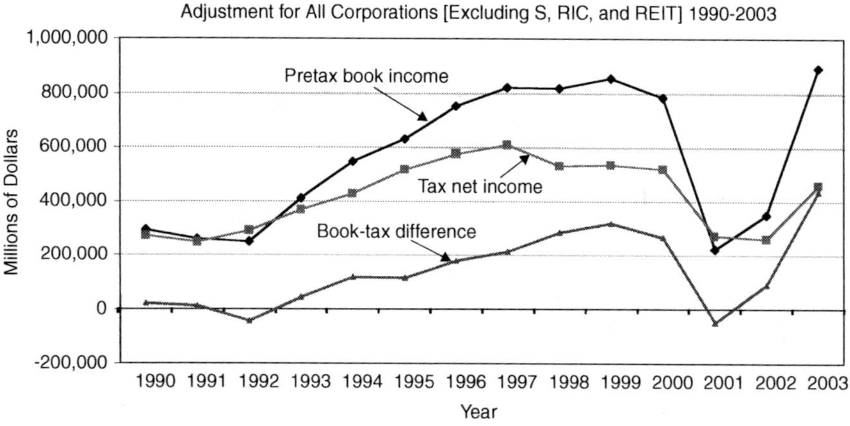

FIGURE 8-1 Pretax book income, tax net income, book-tax difference, M-1 explains, and estimated intercompany dividend.

work by Plesko (2000b, 2002), Plesko and Shumofsky (2005), and Boynton et al. (2005) has established that book-tax differences are not only large, but also growing. Figure 8-2, reproduced from Boynton et al. (2005), shows that book income has differed from taxable income both in scale and in annual changes. Similarly, Table 8-1 shows that these differences are not uniform across industries.

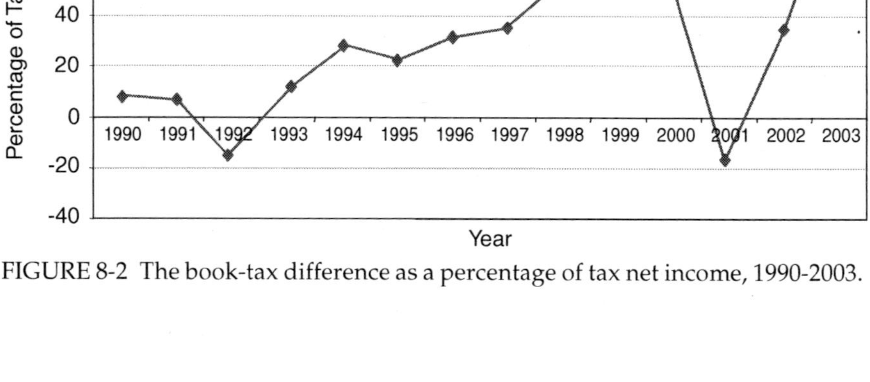

FIGURE 8-2 The book-tax difference as a percentage of tax net income, 1990-2003.

TABLE 8-1 How Significant Are Book-Tax Differences?

	BTD as a Share of Tax Net Income			
Active Corporations	All Industrial Divisions	Raw Materials and Energy Production	Goods Production	Distribution and Transportation of Goods
1995	0.104			
1996	0.140			
1997	0.182			
1998	0.242	–0.088	0.037	0.116
1999	0.690	0.297	0.603	0.471
2000	0.611	0.900	0.406	0.273
2001	–0.150	1.277	–0.026	0.103

As Figure 8-2 shows, there are two years in which the book-tax difference is negative, 1992 and 2001, and both are generally explainable. In 1992, a financial accounting change required firms to begin recognition of expenses to pay postretirement benefits, leading to large reductions in reported book income for that year. The negative difference in 2001 is driven by firms with large negative tax net income that have smaller, or positive, book income. The change in 2003, the most recent year for which data are available, is dramatic, with book income almost double taxable income.

The existence, magnitude, and pattern of book-tax differences raises two important issues for national income purposes: (1) why these differences exist and (2) whether these differences can be identified in financial statements in the years before tax return data become available and corporate profits estimates are finalized. Since both sets of information are driven by the same underlying economic events, the hope is that financial information, which is available sooner than tax information, can complement other available data to make profit estimates and generate both timelier, and more accurate, estimates.

THE TIMING OF INCOME REPORTING

In the case of corporate profits, it is important to note that the amount of information released to investors a short time after the accounting period has ended is large relative to that which can be inferred concurrently

Information	Finance, Insurance, Real Estate, and Rental and Leasing	Professional and Business Services	Education, Health, and Social Assistance	Leisure, Accommodation, and Food Services	Other Services
0.954	0.945	0.064	7.726	0.323	–0.138
1.021	1.356	0.647	7.349	0.097	–0.530
4.830	1.062	0.931	3.286	0.283	–0.503
2.784	0.181	0.272	1.973	0.393	3.047

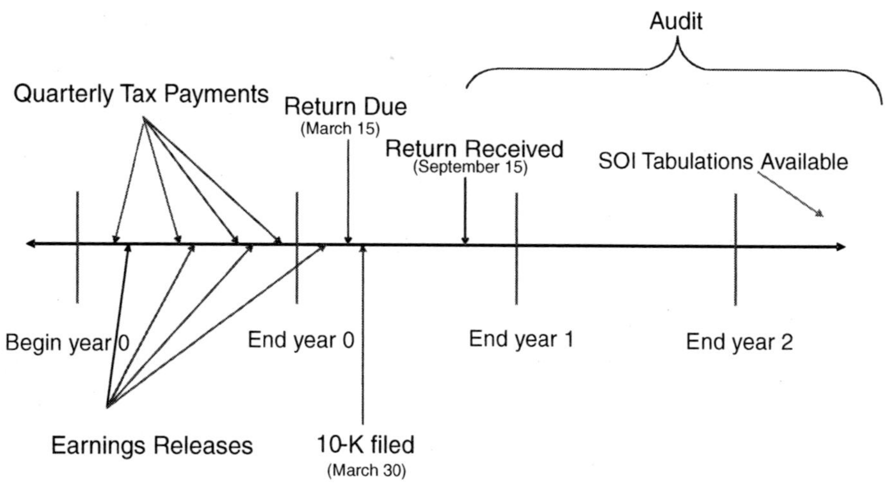

FIGURE 8-3 Timeline of financial and tax reporting.

from the tax system. Figure 8-3 provides an abbreviated time line for a calendar year corporation, that is, one with a year end of December 31.

Companies make quarterly announcements of their profits and quarterly payments of their expected tax liabilities, although these dates do not directly align. Of greatest importance is when annual earnings, which are described here and in Figure 8-3 as year 0, ending on December 31, are disclosed. For publicly traded firms, financial reports are filed by March 30 of the following year. The due date for a calendar year corporation to file a tax return is March 15, prior to the filing of financial statements; however, corporations typically file their returns, after an extension, six months later. As a result, the tax return is not usually filed until September 15.

Shortly after the March 30 filing date, financial data become available in machine-readable form. Tax return data, however, may take two or more years to collect and tabulate and exclude any audit activities. If corporate financial reporting can be used to infer tax attributes, it may be possible to effectively model and estimate taxable income two or more years ahead of having tax return data. Quarterly tax liability patterns are potentially inferable via estimated tax payments; the financial statements, in theory, provide sufficient disclosure of tax items to estimate the taxable income of the firm. In other words, if the financial statement information correctly conveys the information it is intended to, it can be used to predict, or improve the predictions of, the ultimate aggregate tax information eventually reported in the Statistics of Income tabulations.

SOURCES OF DIFFERENCES

Identifying the sources and magnitude of the differences between book income and taxable income has been a growing research interest since the 1999 Treasury Department study. The previous discussion suggests that the differences between book and taxable income are relatively straightforward and dictated by a clear set of regulatory requirements. In practice, however, understanding these differences is not a simple matter. Two different approaches have been used to analyze them, depending on the data available. The first approach uses publicly available data to estimate the book-tax difference using information in financial statements and then attempts to model the amount of book-tax difference as a function of firm and industry characteristics (such as the amount of depreciable assets or foreign operations). The second, direct, approach for analyzing book-tax differences relies on tabulations from tax returns and is discussed by Plesko (2002), Plesko and Shumofsky (2005), and Boynton et al. (2005).

A significant factor in the recent divergence between book and taxable income has been stock options, which do not affect financial accounting earnings but do reduce taxable income. The difficulty in isolating their effect, however, is that, for tax purposes, the option-related expense is included in total compensation and not separately identified. Thus, even with access to tax return data, disclosures may be insufficient to fully understand the causes of the differences.

The new Schedule M-3 (see http://www.irs.gov/pub/irs-utl/2005f1120sm3.pdf) will provide a large amount of additional data to help the IRS, Treasury, and BEA better understand how financial reporting can inform tax administration. Schedule M-3 covers three pages and provides much more information than the Schedule M-1, which consisted of 10 lines.

An important element of the M-3 appears on page 1, where firms are now required to provide identifying information on the financial statement filing entity of which the tax entity is a part. This is followed by a precise derivation of the amount of book income attributable to the tax entity. This reconciliation will better enable tax authorities and analysts to adjust for consolidation differences.

The remaining two pages of the M-3 provide greater detail relative to what was provided prior to 2004, requiring information not only on the amount of an item reported for financial and tax accounting purposes, but also a delineation of the amount of the difference that is temporary and permanent. Among the additional items now reported, stock options appear in Part III on Line 8.

THE BENEFITS OF ADDITIONAL DATA

Two brief examples demonstrate the importance of additional data to analysts. First, in forecasting corporate profits, the additional data provided by the M-3 will offer greater insight into the relation between taxable income and the book profits reported in financial statements. This greater detail should allow for a better use of book income and other information in financial statements to estimate the pattern that taxable income will follow.

Tax policy will also be assisted by the collection of additional information. With the Schedule M-3, not only will analysts have more data concerning the specific operations and organization of a firm, including specific decisions related to tax planning, but also the link to financial statements will make contemporaneous financial information more useful. In estimating the effects of changes in tax policy on businesses, better information about current operations, rather than that reported for previous tax years, should allow for improved estimates of the economic and fiscal effects of proposed changes.

The tax return is the ultimate source of information for determining the effects of tax policy, but it is not the only one. There is substantial evidence suggesting that firms look beyond tax reporting when making tax planning decisions. Changes in corporate behavior, such as investment or financial policy, take place in a tax system that interacts with the capital market through other types of reporting. Without a way to link all of the constraints affecting a business, analysts will have difficulty fully identifying and accurately measuring the effects of changes in tax policy. A better use of both financial and tax information, supplemented by new data provided by the new Schedule M-3, will provide a better understanding of the interrelationships between the two systems.

ACKNOWLEDGMENTS

I have benefited from discussions with Dan Feenberg, Ralph Rector, John Phillips, and David Weber.

REFERENCES

Boynton, C., P. DeFilippes, and E. Legel
2005 Prelude to schedule M-3: Schedule M-1 corporate book-tax difference data 1990-2003. *109 Tax Notes* 1579 (December 19).

Congressional Budget Office
2005 *The Budget and Economic Outlook: An Update.* Washington, DC: U.S. Government Printing Office.

Financial Accounting Standards Board
1992 *Accounting for Income Taxes.* FASB Statement No. 109. Norwalk, CT: Financial Accounting Standards Board.

Hanlon, M.
2003 What can we infer about a firm's taxable income from its financial statements? *National Tax Journal* (4):831-863.

Hanlon, M., S. Kelley LaPlante, and T. Shevlin
2005 *Evidence on the Possible Information Loss of Conforming Book Income and Taxable Income.* Available: http://ssrn.com/abstract=686402 or DOI: 10.2139/ssrn.686402 [accessed May 2006].

Joint Committee on Taxation
2003 *Report of Investigation of Enron Corporation and Related Entities Regarding Federal Tax and Compensation Issues, and Policy Recommendations.* (JCS-3-03). Washington, DC: U.S. Government Printing Office. Available: http://www.gpo.gov/congress/joint/jcs-3-03/vol1/index.html [accessed May 2006].

Manzon, G., and G. Plesko
2002 The relation between financial and tax reporting measures of income. *Tax Law Review* 55(2):175-214.

Mead, C.I., B.R. Moulton, and K. Patrick
2004 *NIPA Corporate Profits and Reported Earnings: A Comparison and Measurement Issues.* Paper presented at the ASSA/AEA meetings, January 3-5, San Diego, CA. Available: http://www.bea.gov/bea/papers.htm [accessed May 2006].

Mills, L.F., and G.A. Plesko
2003 Bridging the reporting gap: A proposal for more informative reconciling of book and tax income. *National Tax Journal* (4):865-893.

Patrick, K.
2001 Comparing NIPA profits with S&P 500 profits. *Survey of Current Business* (April):16-20.

Plesko, G.A.

2000a Book-tax differences and the measurement of corporate income. *Proceedings of the Ninety-second Annual Conference on Taxation* 1999:171-176.

2000b Evidence and theory on corporate tax shelters. *Proceedings of the Ninety-second Annual Conference on Taxation* 1999:367-371.

2002 Reconciling corporation book and tax net income, tax years 1996-1998. *SOI Bulletin* 21(4):1-16.

2003 An evaluation of alternative measures of corporate tax rates. *Journal of Accounting and Economics* 35(2):201-226.

Plesko, G.A., and N. Shumofsky

2005 *Reconciling Corporations' Book and Taxable Income, 1995–2001*. Washington, DC: U.S. Government Printing Office.

Smith, D.T., and Butters, J.K .

1949 *Taxable and Business Income*. Cambridge, MA: National Bureau of Economic Research.

Talisman, J.

2000 Penalty and Interest Provisions, Corporate Tax Shelters. U.S. Department of the Treasury Testimony before the Senate Committee on Finance, March 8.

U.S. Department of the Treasury

1999 *The Problem of Corporate Tax Shelters: Discussion, Analysis and Legislative Proposals.* Washington, DC: U.S. Government Printing Office.